HALONIUM IONS

REACTIVE INTERMEDIATES IN ORGANIC CHEMISTRY

Edited by GEORGE A. OLAH
Case Western Reserve University

A series of collective volumes and monographs on the chemistry of all the important reactive intermediates of organic reactions:

CARBONIUM IONS
Edited by George A. Olah of Case Western Reserve University and Paul v. R. Schleyer of Princeton University: Vol. I (1968), Vol. II (1970), Vol. III (1972). Vol. IV (1973), Vol. V (in press)

RADICAL IONS
Edited by E. T. Kaiser of the University of Chicago and L. Kevan of the University of Kansas (1968)

NITRENES
Edited by W. Lwowski of New Mexico State University (1970)

CARBENES
Edited by Maitland Jones, Jr., of Princeton University and Robert A. Moss of Rutgers University, the State University of New Jersey: Vol. I (1973). Vol. II (1975).

FREE RADICALS
Edited by J. K. Kochi of Indiana University: Vol. I (1973), Vol. II (1973)

HALONIUM IONS
By George A. Olah of Case Western Reserve University (1975).

Planned for the Series

CARBANIONS
By M. Szwarc of the State University of New York, College of Forestry, Syracuse

ARYNES
Edited by M. Stiles of the University of Michigan

HALONIUM IONS

GEORGE A. OLAH
Department of Chemistry
Case Western Reserve University
Cleveland, Ohio

A Wiley-Interscience Publication
JOHN WILEY & SONS New York · London · Sydney · Toronto

Library of Congress Cataloging in Publication Data:

Olah, George Andrew, 1927-
Halonium ions.

(Reactive intermediates in organic chemistry)
"A Wiley-Interscience publication."
Includes bibliographical references and indexes.
1. Halonium ions. I. Title. II. Series.

QD305.H15038 547'.02 75-16417
ISBN 0-471-65329-2

Printed in the United States of America

10 9 8 7 6 5 4 3 2 1

To the memory of Hans Meerwein who more than anybody helped to open the door to the new frontiers of the chemistry of cationic organic intermediates

With valuable contribution by my former coworkers, Dr. Yoke K. Mo (Raychem Corp., Menio Park, California) and Dr. Yorinobu Yamada (Tokyo Institute of Technology) without which this book would not have been realized.

Introduction to the Series

Reactive intermediates have always occupied a place of importance in the spectrum of organic chemistry. They were, however, long considered only as transient species of short life-time. With the increase in chemical sophistication many reactive intermediates have been directly observed, characterized, and even isolated. While the importance of reactive intermediates has never been disputed, they are usually considered from other points of view, primarily relative to possible reaction mechanism pathways based on kinetic, stereochemical, and synthetic chemical evidence. It was felt that it would be of value to initiate a series that would be primarily concerned with the reactive intermediates themselves and their impact and importance in organic chemistry. In each volume, critical, but not necessarily exhaustive coverage is anticipated. The reactive intermediates will be discussed from the points of view of: formation, isolation, physical characterization, and reactions.

The aim, therefore, is to create a forum wherein all the resources at the disposal of experts in the field could be brought together to enable the reader to become acquainted with the reactive intermediates in organic chemistry and their importance.

As the need arises, it is anticipated that supplementary volumes will be published to present new data in this rapidly developing field.

GEORGE A. OLAH

Jim our cocker spaniel may again enlighten for other dog-lovers too this otherwise wasted page.

Preface

Onium ions are well-recognized ionic organic intermediates. They include not only such traditionally established and thoroughly studied classes as ammonium, phosphonium, oxonium, and sulfonium ions, but also higher-valence-state cations of other elements such as carbon and silicon (e.g., carbonium and siliconium ions). Organic halogon cations, that is, halonium ions of an acyclic ($R\overset{+}{X}R$, $Ar\overset{+}{X}R$, and $Ar\overset{+}{X}Ar$) or cyclic ([three-, four- and five-membered cyclic $\overset{+}{X}$ structures]) nature, in recent years have gained increasing significance, both as reaction intermediates and as preparative reagents. They are related to oxonium ions in reactivity but offering greater selectivity.

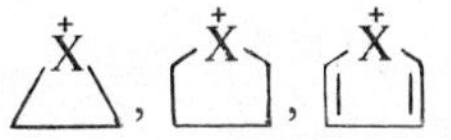

This review has been formulated over several years in connection with the actively pursued research work in our laboratory under the auspices of the National Science Foundation, to whom we are most grateful for support. It concentrates primarily on data relating to halonium ion intermediates as directly observed or isolated species, their structural study, and their chemistry.

It has become increasingly apparent that the field has achieved sufficient significance to warrant a coherent review. Limitations imposed to keep the size of this review to a reasonable length indicated the necessity for dealing only with halonium ions reported as bona fide species and not with the substantial literature concerning the possible involvement of halonium ion intermediates without, however, direct study of the species themselves. In other words, this review is restricted to long-lived organic halonium ions in solution chemistry of as isolated salts. Neither are halogenated carbocations (frequently in equilibrium with bridged halonium ions) discussed, because they are reviewed in the monograph "Carbonium Ions" vol. V. of this series.

It was a pleasure to utilize a quiet and most pleasant stay in the summer of 1974 at the Université Louis Pasteur in Strasbourg, where part of the writing of this book was accomplished. I would like to express my gratitude to my friends in Strasbourg, particularly Professors Guy Ourisson and Jean Sommer, for their outstanding hospitality, and to the university and the Centre National de la Recherche Scientifique for their support. Dr. G. Liang and Mr. S. Prakash are thanked for preparing the indexes, Mrs. M. Hoicourtz and E. Green for typing the manuscript.

GEORGE A. OLAH

Cleavland
June 1975

Contents

1. INTRODUCTION 1

   1.1 General Aspects. 1
   1.2 Classes of Organic Halonium Ions 2
   1.3 Nomenclature 4

Part A. Acyclic Halonium Ions 5

2. ALKYL- AND ARYLHYDRIDOHALONIUM IONS 6

3. DIALKYLHALONIUM IONS 8

   3.1 Preparation and Isolation 9
   3.2 Nmr Spectroscopic Studies 15
      3.2.1 *Pmr* 15
      3.2.2 *Carbon-13 Nmr* 15
   3.3 Raman and Ir Spectroscopic Studies 18
   3.4 Chemical Reactivity 19
      3.4.1 *Alkylation of Aromatics* 19
      3.4.2 *Alkylation of Heteroorganic Compounds* 21
   3.5 Intermolecular Exchange Reactions of Dialkylhalonium Ions with Aryl Halides and Alkylcarbenium Ions 23
      3.5.1 *Determination of Relative Carbocation Stabilities through Competitive Exchange* 28
   3.6 Alkylation and Polymerization of Alkenes 30
   3.7 Role of Dialkylhalonium Ions in Friedel-Crafts Reactions 31
   3.8 Dialkyl alkylenedihalonium Ions 33

4. ALKYLARYLHALONIUM IONS 39

   4.1 Preparation and Nmr Study 39
   4.2 Relative Stability, Alkylating Ability, and Role in Friedel-Crafts Reactions 45
   4.3 Dialkyl phenylenedihalonium Ions 45
   4.4 Trialkyl phenyltriiodonium Ions 52

5. DIAYLHALONIUM IONS 54

   5.1 Diaryliodonium Ions 54
      5.1.1 *Preparation and Isolation* 54

Symmetric Diarylhalonium Ions·Unsymmetric Diarylhalonium Ions·Heteroaryliodonium Ions·Iodonium Ions Bearing Olefinic or Acetylenic Ligands·Di- and Polyiodonium Ions

5.1.2 *Kinetics of Formation* 63
5.1.3 *Structural Aspects* 63

X-Ray Studies·Ir Spectra·Uv Spectra·Nmr Spectra·Substituent Effect of the Phenyliodonio Group·Degree of Dissociation

5.1.4 *Chemical Reactivity* 68

Reactions with Anionic Reagents·Reactions with Amines·Reactions with Arizidines·Reactions with Phosphines·Reactions with Alkyl Phosphites and Phosphates·Reactions with Mercury·O-Phenylation Reactions·Reactions with Inorganic Nucleophiles·Reactions of Heteroaryliodonium Ions

5.1.5 *Biological Activity* 87
5.2 Diarylbromonium and -chloronium Ions 88

6. HALONIUM YLIDES 90

6.1 Alkylhalonium Ylides 90
6.2 Arylhalonium Ylides 90

Part B. Cyclic Halonium Ions

7. ETHYLENEHALONIUM IONS 98

7.1 Preparation by Neighboring Halogen Participation and Pmr Study 98
7.1.1 *Parent Ethylenehalonium Ions* 98
7.1.2 *Propylenehalonium Ions* 101
7.1.3 *2,3-Dimethylethylenehalonium Ions* 103
7.1.4 *2,2-Dimethylethylenehalonium Ions* 106
7.1.5 *Trimethylethylenehalonium Ions* 107
7.1.6 *Tetramethylethylenehalonium Ions* 108
7.1.7 *1-Butyleneiodonium Ion (2-Ethylethyleneiodonium Ion)* 111
7.1.8 *Propadienylhalonium Ions* 113
7.2 Carbon-13 Nmr Study of Ethylenehalonium Ions 113
7.2.1 *Symmetrically Substituted Ethylenehalonium Ions* 115
7.2.2 *Unsymmetrically Substituted Eithylenehalonium Ions* 116
7.3 Preparation via Protonation of Cyclopropyl Halides 118
7.4 Preparation via Direct Halogenation of Olefins 120
7.5 Differentiation of π- and σ-Bonded Halogen Complexes of Alkenes 120

8. Tetramethylenehalonium Ions 125

8.1 Preparation by Halogen Participation and Pmr Study 125
8.1.1 *Parent Tetramethylenehalonium Ions* 125
8.1.2 *2-Methyltetramethylenehalonium Ions* 128
8.1.3 *2,5-Dimethyltetramethylenehalonium Ions* 129
8.1.4 *Tri- and Tetramethyltetramethylenehalonium Ions* 133
8.1.5 *Halogenated Tetramethylenehalonium Ions* 134
8.1.6 *2-Methylenetetramethylenehalonium Ions* 134
8.2 Carbon-13 Nmr Study of Tetramethylenehalonium Ions 135
8.2.1 *Symmetrically Substituted Ions* 135
8.2.2 *Unsymmetrically Substituted Ions* 138

9. Tri- and Pentamethylenehalonium Ions 140

9.1 Trimethylenehalonium Ions 140
9.2 Pentamethylenehalonium Ions 142

10. Bicyclic Halonium Ions 144

10.1 Cyclopentenebromonium Ion (2-Bromonia[3.1.0] bicyclohexane) 144
10.2 Chloronia[4.3.0] bicyclononane 146

11. Stability and Chemical Reactivity of Alicyclic Halonium Ions 147

11.1 Stability of Cyclic Halonium Ions 147
11.2 The Role of σ and π Complexes in the Addition of Halogen to Olefins and Acetylenes 148
11.2.1 *Alkenes* 149
11.2.2 *Alkynes* 154
11.3 Reactions of Ethylenehalonium Ions 155
11.4 Reactions of Tetramethylenehalonium Ions 156

12. Heteroaromatic Halophenium Ions 158

12.1 Halophenium Ions 158
12.2 Benziodophenium Ions 158
12.3 Benzchlorophenium Ions 161
12.4 Dibenzhalophenium Ions 163

13. Conclusions and Future Outlook 169

References 170

Author Index 177

Subject Index 181

HALONIUM IONS

CHAPTER 1

Introduction

1.1 GENERAL ASPECTS

Organic halonium ions, as we now recognize them, were first prepared in case of some remarkable stable diaryliodonium compounds. In 1894, Hartmann and Meyer[1] were the first to prepare a diphenyliodonium ion salt when they reacted iodosobenzene in concentrated H_2SO_4 and obtained *p*-iodophenyl, phenyliodonium bisulfate. Diphenyliodonium ions have since been studied by various research groups, most notably since 1950 by Nesmeyanov[2] by Beringer. Diarylbromonium and -chloronium ions, although considerably less stable and much less investigated, were also prepared in the 1950s by Nesmeyanov and coworkers[2]. As in the case of triarylcarbenium ions, these aryl-substituted ions were considered to be of exceptional stability and the only existing type of organic halonium ion; thus they were thought to have no particular significance or revelance otherwise in organic chemistry.

Open-chain dialkylhalonium ions of the type R_2X^+ (X = I, Br, Cl) were unknown until recently, as were alkylarylhalonium ions ($Ar\overset{+}{X}R$). Realization of their possible role as intermediates in alkylation reactions of haloalkanes and -arenes has followed their recent preparation and study.[3]

One of the most daring proposals of a organic reaction mechanism of its time was made in 1937 by Roberts and Kimball[4] who suggested that the observed trans stereospecificity of the bromine addition to alkenes is a consequence of intermediate bridged bromonium ion formation. The brief original publication suggested the the actual structure of the ion is undoubtedly intermediate between **1** and **2**. Structure **1** was not intended to represent a conventional free carbenium ion, however. Since the two carbons in either

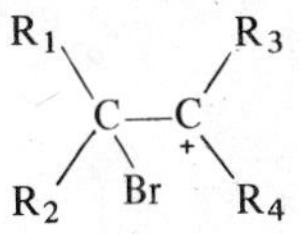

R1 R3 C—C Br + R2 R4

1

R1 R3 C—C +Br R2 R4

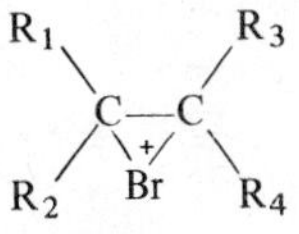

2

structure are joined by a single bond and by a halogen bridge, free rotation is not to be expected. A clear description of the difference in binding between carbon and bromine in **1** and **2** was not given.

The bromonium ion concept was quickly used by other investigators to account for stereospecific transformation of olefins,[5] notably by Winstein and Lucas,[6] but was not unanimously accepted.[7] For example, in discussing the mechanistic concepts of bromonium ion formation, Gould,[8] in his still popular text, wrote in 1959, "Although a number of additions are discussed in terms of the halogenonium-ion mechanism, the reader should bear in mind that few organic mechanisms have been accepted so widely while supported with such limited data."

In 1967, Olah and Bollinger[9] reported the first preparation and spectroscopic characterization of stable, long-lived bridged alkylenehalonium ions. This was followed by Olah and DeMember's[3] preparation in 1969 of the first dialkylhalonium ions. Since then the field of organic halonium ions has undergone rapid development through substantial contributions from an increasing number of investigators, notably by Peterson.[10]

The study of organic halonium ions has clearly reached a point where a comprehensive survey was needed. This review considers halonium ions reported as intermediate, directly observed, or isolated species, but does not discusses in detail halogen additions and related reactions in which halonium ion intermediates could have been involved but were not directly studied. Organic halonium ions can be defined as organic ligand substituted halogen cations (generally divalent, as no examples of higher-valence ions, although possible, have been reported) and are recognized as an important class of onium ions. The halogen atom is organic halonium ions is generally bound to two carbon atoms [although in the case of acidic halonium ions, that is, protonated alkyl and aryl halides (RX^+H) one ligand may be hydrogen].

1.2 CLASSES OF ORGANIC HALONIUM IONS

Organic halonium ions can be divided into two main categories: (1) acyclic (open-chain) halonium ions and (2) cyclic halonium ions.

In the first class, diarylhalonium (ArX^+Ar),[11] alkylarylhalonium ($R\overset{+}{X}Ar$),[12] and diakylhalonium ($R_2\overset{+}{X}$)[3] ions are known. Recently, open-chain di- and trihalonium ions have also been prepared.[13]

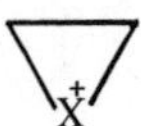

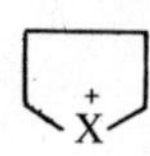

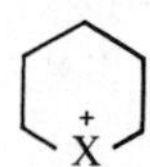

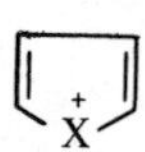

In the second class, three-membered-ring ethylenehalonium,[9,14] five-membered-ring tetramethylene halonium,[15] and six-membered-ring pentamethylenehalonium[16] ions are best known, although some four-membered-ring ions have also been reported.[17,18] Substantial interest[19] has also been shown in heteroaromatic halophenium ions.

1.3 NOMENCLATURE

The naming of organic cations is not yet governed by definite rules of the International Union of Pure and Applied Chemistry (IUPAC). In most publications in the field, and in this review, the trivial naming system adopted by analogy with other onium ions is used: alkylenehalonium ions for cyclic, three-membered-ring ions of the type

$$\underset{\diagdown \overset{+}{X} \diagup}{R_2C \text{—} CR_2};$$

tri-, tetra-, and pentamethylenehalonium ions for four-, five-, and six-membered-ring cyclic halonium ions; dialkyl-, alkylaryl-, and diarylhalonium ions for open-chain halonium ions of the R_2X^+, $RArX^+$, and Ar_2X^+ type, respectively. Acidic halonium ions of the $R\overset{+}{X}H$ or $Ar\overset{+}{X}H$ type are named as hydrido alkyl- or arylhalonium ions or as protonated haloalkanes or -arenes, respectively.

With the proliferation of ring sizes among cyclic halonium ions, and the isolation of recrystallized salts, Peterson[16] felt it appropriate to discuss nomenclature systems for this new class of heterocycle. He consulted with the Chemical Abstracts Service and came to the following conclusions. The names for polymethylenehalonium ions used by Olah and by Peterson are approved IUPAC names. However, an extension of the Hantsch-Widman[20] names for heterocycles (see Table 1) often leads to shorter names and has already been the basis for names appearing in *Chemical Abstracts.* A third system of nomenclature is based on the substitution of the positive halogen for CH_2 in the appropriate cycloalkane (replacement names).[21] Such names are used for complex heterocycles (such as bicyclic compounds), of which at least two examples have been reported in the case of halonium ions: 3-chloronia[4.3.2.]bicyclononane[22] and 2-bromonia[3.1.0]bicyclohexane.[23]

Unfortunately, different numbering systems apply for the different nomenclatures. In the Hantsch-Widman and replacement systems for monocyclic rings, the halogen is numbered 1. (Note that for polycyclic systems the halogen is not necessarily numbered 1 in the replacement system.) In the polymethylene system, carbon was numbered 1 in the early papers on ethylenebromonium ions, while halogen was numbered 1 for tetramethylenehalonium ions. It now seems reasonable that nomenclature and numbering should be made uniform. In the present review, we use exclusively the polymethylene system with halogen numbered 1, with the exception of bicyclic halonium ions in which the systematic replacement nomenclature is used.

TABLE 1

Nomenclature of Cyclic Halonium Ions

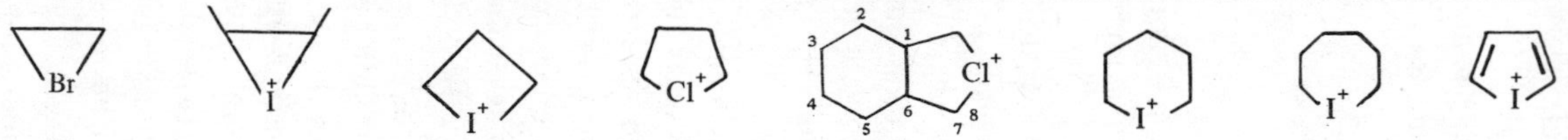

Compound	Polymethylene or heteroarene name[a]	Hantasch-Widman name[b]	Replacement name[c]
a	Ethylenebromonium	Bromiranium[d]	Bromoniacyclopropane
b	2,3-Dimethylethyleneiodonium	2,3-Dimethyliodiranium[d]	2,3-Dimethyliodoniacyclopropane
c	Trimethyleneiodonium	Iodetanium	Iodoniacyclobutane
d	Tetramethylenechloronium	Chlorolanium[d]	Chloroniacyclopentane
e	*cis*-1,2-Cyclohexylenedimethylene-chloronium	*cis*-Hexahydro-2-benzo-chlorolanium	8-Chloronia-*cis*-bicyclo[4.3.0]nonane
f	Pentamethyleneiodonium	Iodanium[e]	Iodoniacyclohexane
g	Hexamethyleneiodonium	Iodepanium	Iodoniacycloheptane
h	Iodophenium		Iodoniacyclopentadiene

[a] IUPAC, *Definitive Rules for Nomenclature of Organic Chemistry*, Section C, Butterworths, London, 1965, Rules C-82, C-107.

[b] IUPAC, *Definitive Rules for Nomenclature of Organic Chemistry*, Section A and B, Butterworths, London, 1966, Rule D; see also *The Naming and Indexing of Chemical Compounds from Chemical Abstracts*, American Chemical Society, Washington, D.C., 1957, Rule 375.

[c] Clossary section of IUPAC publication referred to in footnotes a and b, Rules B-4 and C-82.3.

[d] Name appearing in *Chem. Abstr., Subject Index*, **69**, (1968).

[e] Kurt L. Loening, director of nomenclature of the Chemical Abstracts Service, has suggested that the name might be misleading, but Peterson has chosen the shorter name for the sake of simplicity (see ref. 16).

PART A

Acyclic Halonium Ions

CHAPTER 2

Alkyl- and Arylhydridohalonium Ions

The self-condensation of alkyl halides in superacid solutions represents a convenient synthetic route to symmetric dialkylhalonium ions ($R\overset{+}{X}R$) (see chapter 3). This reaction formally corresponds to the acid-catalyzed condensation of alcohols to ethers and, by analogy, involves hydridohalonium ions ($R\overset{+}{X}H$) as intermediates which subsequently undergo nucleophilic attack by excess alkyl halide.

$$CH_3X \underset{}{\overset{H^+}{\rightleftharpoons}} CH_3\overset{+}{X}H \longrightarrow CH_3\overset{+}{X}CH_3 + HI$$
$$X = Cl,\ Br,\ I \qquad :\ddot{X}-CH_3$$

When a solution of iodomethane in SO_2ClF is added to a solution of HSO_3F-SbF_5 in SO_2ClF at $-78°$, the pmr and cmr spectra both exhibit two resonances, substantially deshielded from methyl ioditic, in the ratio 2:1.

$$CH_3X \qquad CH_3\overset{+}{X}CH_3 \qquad CH_3\overset{+}{X}H$$
$$X = Cl,\ Br,\ I \qquad X = Cl,\ Br,\ I \qquad X = Br,\ I$$

No unreacted iodomethane can be detected. The major species (cmr $\delta_{^{13}C}$ 8.7, pmr δ 3.56) formed is the dimethyliodonium ion, while the minor species (cmr $\delta_{^{13}C}$ 2.7, pmr δ 4.10) is the methylhydridoiodonium ion;[24] the carbon shielding in the methylhydridoiodonium ion compared to that in the dimethyliodonium ion is consistent with the removal of a β-CH_3 group, and both become quartets in the proton-coupled cmr spectra with $^1J_{CH}$ values substantially larger than that of iodomethane (Table 2). Bromomethane under the same conditions yields two carbon resonances in the ratio 8:1. The major species (δ 37.5) is the dimethylbromonium ion, while the minor species (δ 32.6) is the methylhydridobromonium ion; no unreacted bromomethane can be detected. Chloromethane reacts under the same conditions to yield only dimethylchloronium ions (cmr δ 49.9) and unreacted chloromethane (cmr δ 26.0).

Halobenzenes do not form diphenylhalonium ions under superacidic conditions, but show that the reaction competing with halogen protonation is ring protonation to benzenium ions.[25] Indeed, chlorobenzene and bromobenzene

$$C_6H_5I \xrightarrow[SO_2ClF,\ -78^\circ]{FSO_3H\text{-}SbF_5} C_6H_5\overset{+}{I}H$$

quantitatively yield the corresponding 4-halobenzenium ions on protonation with FSO_3H-SbF_5-SO_2ClF at -78°. Iodobenzene under the same conditions yields a single ion with deshielded pmr and cmr resonances similar to those of the methyl phenyliodonium ion (Table 1). The iodine-protonated hydridohalonium ion structure is therefore assigned to this ion; the shielding of C_{ipso} and the deshielding of C_{ortho} in CH_5IH^+ compared to $C_6H_5\overset{+}{I}CH_3$ is consistent with the removal of a β- and a γ-CH_3 group, respectively. In the case of $CH_3\overset{+}{I}H$, the proton in $C_6H_5\overset{+}{I}H$ evidently exchanges rapidly (on the nmr scale) with the excess acid. $C_6H_5\overset{+}{I}H$ does not, however, rearrange to ring-protonated $C_6H_6I^+$, even when the temperature is raised to -20°

TABLE 2
Cmr Data for Alkyl- and Arylhydridohalonium Ions, Their Precursors, and Related Model Ions

Compound	Cmr data[a]
CH_3I[b]	CH_3, -21.5; $J_{CH} = 150.3$
$CH_3I^+CH_3$[c]	CH_3, 8.7; $J_{CH} = 158.1$
CH_3I^+H[c]	CH_3, 2.7; $J_{CH} = 155.2$
CH_3Br[b]	CH_3, 10.8
$CH_3Br^+CH_3$[c]	CH_3, 37.5
CH_3Br^+H[c]	CH_3, 32.6
C_6H_5I[b]	C_{ipso}, 95.1; C_{ortho}, 137.9; C_{meta}, 130.8; C_{para}, 128.0
$C_6H_5I^+H$[c]	C_{ipso}, 100.0; C_{ortho}, 138.6; C_{meta}, 132.8; C_{para} 132.8
$C_6H_5I^+CH_3$[b]	C_{ipso}, 105.2; C_{ortho}, 137.6; C_{meta}, 133.7; C_{para} 133.7

[a]Chemical shifts are in parts per million from external capillary trimethylsilane (TMS). Coupling constants are in hertz.
[b]In SO_2ClF at -70°.
[c]In FSO_3H-SbF_5-SO_2ClF at -70°.

CHAPTER 3

Dialkylhalonium Ions

Dialkylhalonium ions were first obtained as stable fluoroantimonates and characterized by Olah and DeMember[3] in 1969. Subsequently, the isolation and carbon-13 nmr, Raman, and ir spectroscopic study of a series of dialkylhalonium ions, including the parent dimethylhalonium ions, was reported.[26] The alkylating ability[26] as well as the intermolecular exchange reactions[27] of dilakylhalonium ions were also studied.

3.1 PREPARATION AND ISOLATION

There are two general methods for the preparation of dialkylhalonium ions: (1) the reaction of excess primary and secondary alkyl halides with SbF_5-SO_2, anhydrous fluoroantimonic acid (HF-SbF_5), or anhydrous silver hexafluoroantimonate (or related complex fluoro silver slats) in SO_2 solution, and (2) the alkylation of alkyl halides with methyl or ethyl fluoroantimonate (or alkylcarbenium fluoroantimontes) in SO_2 solution.[3,26] The first method is

$$2RX + SbF_5\text{-}SO_2 \xrightarrow[SO_2]{-60^\circ} R\overset{+}{X}R\ SbF_5X^- \quad (1)$$

$$\text{or } 2RX + HSbF_6 \rightarrow R\overset{+}{X}R\ SbF_6^- + HX$$

$$\text{or } 2RX + AgSbF_6^- \rightarrow R\overset{+}{X}R\ SbF_6^- + AgX$$

$$R = CH_3,\ C_2H_5,\ i\text{-}C_3H_7$$

$$X = Cl,\ Br,\ I$$

$$R'X + R^+SbF_6^- \xrightarrow[-60^\circ]{SO_2} R'\overset{+}{X}R\ SbF_6^- \quad (2)$$

$$R' = CH_3,\ C_2H_5,\ C_3H_7$$

$$R = CH_3,\ C_2H_5$$

$$X = Cl,\ Br,\ I$$

only suitable for the preparation of symmetric dialkylhalonium ions, while the second method can be used for both symmetric and unsymmetric dialkylhalonium ions.

Additional methods for the preparation of dilakylhalonium ions [equations (3) to (7)] were also found. However, these methods generally have less practical value because they usually yield products contaminated by other ions in the same solution. (3) The reaction of alkyl fluoroantimonates with metal halides in SO_2 solution gives the corresponding dialkylhalonium ions.

$$2R^+SbF_6^- + MX \xrightarrow{SO_2} RX^+R\ SbF_6^- + MSbF_6^- \qquad (3)$$

$$R = CH_3,\ C_2H_5,\ C_3H_7$$

$$M = Na,\ K,\ etc.$$

$$X = Cl,\ Br,\ I$$

(4) Tertiary alkyl halides react with alkyl fluoroantimonates to give the corresponding tertiary alkylcarbenium ions and the symmetric dialkylhalonium

$$R^+SbF_6^- + R_3CX \xrightarrow{SO_2} RX^+CR_3\ SbF_6^- \xrightarrow{R^+SbF_6^-} RX^+R\ SbF_6^- + R_3C^+\ SbF_6^- \qquad (4)$$

ions. (5) When an excess of alkyl fluoroantimonate was treated with dihalides such as 1,4-dihalobutanes, the initially formed dialkylalkylenedihalonium ions were cleaved to give five-membered-ring halonium ions and dialkylhalonium

$$2R^+SbF_6^- + X(CH_2)_4X \rightarrow [RX^+(CH_2)_4\overset{+}{X}R\ \ 2SbF_6^-] \rightarrow RX^+R\ SbF_6^- + \text{(five-membered ring } \overset{+}{X}\text{)}\ SbF_6^- \qquad (5)$$

ions.[28] (6) When haloalkylcarbenium ions, such as dimethylchlorocarbenium ions, were treated with the analogous alkanes, rapidly exchanging diisopropylchloronium ions were obtained[27] (see Section 3.5).

$$CH_3\overset{+}{C}ClCH_3 + CH_3CH_2CH_3 \rightleftharpoons CH_3CHClCH_3 + CH_3C^+HCH_3 \rightleftharpoons [(CH_3)_2CH]_2\overset{+}{Cl} \qquad (6)$$

chloronium ions were obtained[27] (see Section 3.5). (7) Dialkylhalonium ions can also be directly obtained from alkanes and SbF_5-Cl_2-SO_2ClF solution at $-78°$.[29] For example, ethane reacted with this powerful chlorinating reagent to give the dimethylchloronium ion.

$$CH_3CH_3 + SbF_5\text{-}Cl_2\text{-}SO_2ClF \rightarrow CH_3Cl^+CH_3\ SbF_5Cl^- \qquad (7)$$

The details of the preparation of a series of dialkylhalonium ions are summarized in Table 3. Generally, the preparation of symmetric dialkylhalonium ions is more simple and the reactions are clean. Unsymmetric dialkylhalonium ions undergo disproportionation and alkylation (self-condensation) reactions, even at low temperatures (ca. $-30°$). Data in Table 2 (see remarks) also suggest a qualitative measure of the decreasing stability of unsymmetric halonium ions in the order $\overset{+}{I} > \overset{+}{Br} > \overset{+}{Cl}$ and an increasing tendency for disproportionation according to $\overset{+}{Cl} > \overset{+}{Br} > \overset{+}{I}$.

Dimethylhalonium fluoroantimonates such as dimethylbromonium and -iodonium fluoroantinomates can be isolated as crystalline salts. They are stable in a dry atmosphere at room temperature, and some are now commercially available. Dimethylhalonium fluoroantimonate salts are very hygroscopic, and exposure to atmospheric moisture leads to immediate hydrolysis.

TABLE 3
Preparation of Dialkylhalonium Ions

Reagents and temperature (°C)	Method of preparation	Dialkylhalonium ion formed	Remarks
$2CH_3X$ (X = Cl, Br, I), SbF_5-SO_2, -60°	(1)		Preferred preparative method
CH_3X, (X = Cl, Br, I) CH_3F-SbF_5-SO_2, -60°	(2)	$CH_3X^+CH_3$ SbF_5X-	Preparative method
MX, (M = Na, K, etc.; X = Cl, Br, I) $2CH_3F$-SbF_5-SO_2, -60°	(3)		Less practical
R_3CX, CH_3F-SbF_5-SO_2, -60°	(4)		Formation of R_3C^+
$X(CH_2)_nX$ (X = Cl, Br; n = 4, 5) CH_3F-SbF_5SO_2	(5)		Formation of five-membered-ring halonium ions
CH_4 or C_2H_6, SbF_5-Cl_2-SO_2ClF, -78°	(7)		Poor yield
CH_3CH_2X, (X = Cl, Br, I) CH_3F-SbF_5-SO_2, -78°	(2)	$CH_3X^+CH_2CH_3$ SbF_5X^- ↓ X = Cl (50%) Br (90%), I (100%)	Preferred method
CH_3X, (X = Cl, Br, I) CH_3CH_2F-SbF_5SO_2, -78°	(2)	$CH_3X^+CH_3$ SbF_5X^- + $(CH_3CH_2)_2X^+$ SbF_5X^- X= Cl (50%), Br (10%), I (0%)	
$2CH_3CH_2X$, (X = Cl, Br, I) SbF_5-SO_2, -78°	(1)		Preferred preparative method
CH_3CH_2X, (X = Cl, Br, I) CH_3CH_2F-SbF_5-SO_2	(2)	$(CH_3CH_2)_2X^+$ SbF_5X^-	Preparative method

MX (M = Na, K, etc.; X = Cl, Br, I), $2CH_3CH_2F$-SbF_5-SO_2, -78°	(3)		Less practical
R_3CX, CH_3CH_2F-SbF_5-SO_2 -78°	(4)		Formation of R_3C^+
$CH_3CH_2CH_2X$ (X = Cl, Br, I) CH_3F-SbF_5-SO_2, -78°	(2)	$CH_3X^+CH_2CH_2CH_3\ SbF_6^-$ X = Cl (not observed), Br (10%), I (100%) $CH_3X^+CH(CH_3)_2$ + $CH_3X^+CH_3\ SbF_5X^-$ SbF_5X^- X = Cl (100%), Br (90%), I (0%)	These halonium ions are not stable at -30° and disproportionate to dimethylhalonium ions and some alkylation products.[a] Iodonium ion is stable at -40°
$(CH_3)_2CHX$ (X = Cl, Br, I), CH_3F-SbF_5-SO_2, -78°	(2)	$CH_3X^+CH(CH_3)_2\ SbF_5X^-$ X = Cl (major), Br (major), I (major) $[(CH_3)_2CH]_2X^+\ SbF_5X^-$ + $CH_3X^+CH_3\ SbF_5X^-$ X = Cl (minor), Br (minor), I (minor); X = Cl, Br, I	These ions are not stable at -30° and decompose to alkylation products
$CH_3CH_2CH_2X$ (X = Cl, Br, I), CH_3CH_2F-SbF_5-SO_2, -78°	(2)	$CH_3CH_2X^+CH_2CH_2CH_3\ SbF_5X^-$ X = Cl (not observed), Br, I $\downarrow$ CH_3CH_2F-SbF_5, -30° $(CH_3CH_2)_2X^+\ SbF_5X^-$ + alkylation products[a] X = Cl, Br	Iodonium ion is stable at -20°

TABLE 3
Preparation of Dialkylhalonium Ions (Continued)

Reagents and temperature (°C)	Method of preparation	Dialkylhalonium ion formed	Remarks
$(CH_3)_2CHX$ (X = Cl, Br, I), CH_3CH_2F-SbF_5-SO_2, −78°	(2)	$CH_3CH_2X^+CH(CH_3)_2\ SbF_5X^-$ X = Cl, Br, I; −20° ↓ $(CH_3CH_2)_2X^+$ + alkylation products[a]	
$2CH_3CH_2CH_2X$) (X = Cl, Br, I), SbF_5-SO_2, −78°	(1)	$(CH_3CH_2CH_2)_2X^+\ SbF_5X^-$ X = Cl (0%), Br (15%), I (15%), $[(CH_3)_2CH]_2X^+\ SbF_5X^-$ X = Cl, (100%), Br (85%), I (15%)	Exchanging dialkylhalonium ions
$2(CH_3)_2CHX$, (X = Cl, Br, I) SbF_5-SO_2, −78°	(1)	$[(CH_3)_2CH]_2X^+\ SbF_5X^-$	Exchanging dialkylhalonium ions
$2(CH_3)_3CX$ (X = Cl, Br, I), SbF-SO_2, −78°	(1)	$[(CH_3)_3C]_2X^+\ SbF_5X^-$	Exchanging dialkylhalonium ions

[a] For example, formation of *tert*-hexyl cation.

TABLE 4

Pmr Parameters of Dialkylhalonium Ions[a]

Ion	Method of preparation	CH_3X^+	$-CH_2X^+-$	$-CHX^+-$	CH_3CX^+-	$-CH_2CX^+-$	CH_3CCX^+
$CH_3Cl^+CH_3$	(1)	4.20 (s)					
$CH_3Br^+CH_3$	(1)	4.13 (s)					
$CH_3I^+CH_3$	(1)	3.60 (s)					
$CH_3Cl^+CH_2CH_3$	(2)	4.41 (s)	5.23 (q, J = 7)		1.98 (t, J = 7)		
$CH_3Br^+CH_2CH_3$	(2)	4.09 (s)	5.10 (q, J = 7)		2.14 (t, J = 7)		
$CH_3I^+CH_2CH_3$	(2)	3.57 (s)	4.60 (q, J = 7)		2.17 (t, J = 7)		
$(CH_3CH_2)_2Cl^+$	(1)		5.20 (q, J = 7)		1.93 (t, J = 7)		
$(CH_3CH_2)_2Br^+$	(1)		5.20 (q, J = 7)		2.20 (t, J = 7)		
$(CH_3CH_2)_2I^+$	(1)		4.74 (q, J = 7.5)		2.30 (t, J = 7.5)		
$CH_3Br^+CH_2CH_2CH_3$	(2)	4.21 (s)	5.11 (t, J = 7)			2.3 (m)	1.34 (t, J = 7)
$CH_3I^+CH_2CH_2CH_3$	(2)	3.60 (s)	4.70 (t, J = 7)			2.33 (sex, J = 7)	1.33 (t, J = 7)
$CH_3Cl^+CH(CH_3)_2$	(2)	4.48 (s)		6.20 (hep, J = 6)	2.18 (d, J = 6)		
$CH_3Br^+CH(CH_3)_2$	(2)	4.18 (s)		6.21 (hep, J = 6)	2.23 (d, J = 6)		
$CH_3I^+CH(CH_3)_2$	(2)	3.69 (s)		5.90 (hep, J = 6.5)	2.40 (d, J = 6.5)		
$CH_3CH_2Br^+CH_2CH_2CH_3$	(2)		5.13 (q, J = 7)		2.30 (t, J = 7)		
			5.13 (t, J = 7)			2.32 (m)	1.30 (t, J = 7)
$CH_3CH_2I^+CH_2CH_2CH_3$	(2)		4.67 (q, J = 7)		2.24 (t, J = 7)		
			4.70 (t, J = 7)			2.38 (sex, J = 7)	1.30 (t, J = 7)
$CH_3CH_2Cl^+CH(CH_3)_2$	(2)		5.10 (q, J = 7.5)	6.23 (hep, J = 6)	2.03 (t, J = 7.5)		
					2.19 (d, J = 6)		
$CH_3CH_2Br^+CH(CH_3)_2$	(2)		4.94 (q, J = 7)	6.21 (hep, J = 7)	2.09 (t, J = 7)		
					2.25 (d, J = 7)		

TABLE 4

Pmr Parameters of Dialkylhalonium Ions[a] (continued)

Ion	Method of preparation	CH_3X^+	$-CH_2X^+-$	$-CHX^+-$	CH_3CX^+-	$-CH_2CX^+-$	CH_3CCX^+
$CH_3CH_2I^+CH(CH_3)_2$	(2)		4.70 (q, J = 7.5)	5.90 (hep, J = 7)	2.30 (t, J = 7) 2.37 (d, J = 7)		
$(CH_3CH_2CH_2)_2Br^+$	(1)		5.23 (t, J = 7)			2.3 (m)	1.31 (t, J = 7)
$(CH_3CH_2CH_2)_2I^+$	(1)		4.61 (t, J = 7)			2.4 (m)	1.18 (t, J = 7)
$[(CH_3)_2CH]_2Br^+$	(1)			6.30 (hep, J = 7)	2.30 (d, J = 7)		
$[(CH_3)_2CH]_2I^+$	(1)			5.84 (hep, J = 7)	2.10 (d, J = 7)		

[a]From external capillary TMS. Spectra were recorded in SO_2 solution at 60 MHz. Some of the pmr shifts of dialkylhalonium ions differ uniformly by 0.4-0.6 ppm from those reported in ref. 31. s, Singlet; d, doublet; t, triplet; q, quartet; sex, sextet; hep, heptet; m multiplet.

3.2 NMR SPECTROSCOPIC STUDIES

3.2.1 Pmr

The pmr parameters of studied dialkylhalonium ions are summarized in Table 4. The methyl proton shifts of all the methylated halonium ions are deshielded in the order $\diagup\overset{+}{Cl}\diagdown > \diagup\overset{+}{Br}\diagdown > \diagup\overset{+}{I}\diagdown$, indicating the inductive effective of the halogen atoms. Chlorine, being the smallest halogen atom in halonium ions (fluoronium ions are not known in solution) can accommodate the least the charge, whereas iodine, the largest of the halogen atoms, can accept essentially most of the charge. The deshielding effect of the methyl protons in dimethylhalonium ions is also related to the inductive order of the halogen atoms. The α-methylene ($-CH_2X^+-$) and α-methine ($-CHX^+-$) protons of homologous dialkylhalonium ions show a similar order of deshielding ($\diagup\overset{+}{Cl}\diagdown > \diagup\overset{+}{Br}\diagdown > \diagup\overset{+}{I}\diagdown$). However, the β-methyl protons of related homologous dialkylhalonium ions show an opposite deshielding order ($\diagup\overset{+}{I}\diagdown > \diagup\overset{+}{B}\diagdown > \diagup\overset{+}{Cl}\diagdown$), indicating that the inductive effect of the positively charged halogen atoms is diminishing and the anisotropy effect of the halogen atoms is causing an opposite trend ($\diagup\overset{+}{I}\diagdown > \diagup\overset{+}{Br}\diagdown > \diagup\overset{+}{Cl}\diagdown$).

3.2.2 Carbon-13 Nmr

Olah and DeMember[26a] reported the first carbon-13 nmr study of dimethylhalonium ions. Later Fourier-transform carbon-13 nmr allowed a more thorough study of dialkylhalonium ions.[26c,d]

The carbon-13 chemical shifts of dialkylhalonium ions obtained by the Fourier-transform method are summarized in Table 5. The assignments were

TABLE 5
Carbon-13 Chemical Shifts of Dialkylhalonium Ions (RX^+R')[a]

			R′			R	
R	R′	X	α-C	β-C	γ-C	α-C	β-C
CH_3	CH_3	Cl	48.9			48.9	
CH_3	CH_3	Br	37.6			37.6	
CH_3	CH_3	I	9.5			9.5	
CH_3	C_2H_5	Cl	73.5	16.9		48.0	
CH_3	C_2H_5	Br	66.9	16.7		36.1	
CH_3	C_2H_5	I	37.9	17.3		8.0	
C_2H_5	C_2H_5	Cl	73.1	20.5		73.1	20.5
C_2H_5	C_2H_5	Br	64.4	18.1		64.4	18.1
C_2H_5	C_2H_5	I	36.9	18.1		35.9	18.1
CH_3	n-C_3H_7	I	48.9	26.3	17.1	9.5	

[a]In parts per million from TMS. In SO_2 at −30 to −40°.

made by the usual procedures of Grant and coworkers.[30] These include the use of "off-resonance" proton decoupling, as well as considerations of relative signal intensities and molecular symmetry. Also, the observation that polar groups exert a large inductive effect on the shifts of directly attached carbons was taken into consideration.

A comparison of the carbon shifts in dialkylhalonium ions and alkanes (Table 6) shows the effect of the RX^+ group on carbon shieldings along a hydrocarbon chain. It can be seen from the available data that RX^+ has an appreciable deshielding effect at the α and β carbons. The α effect correlates with the electronegativities of the halogens in that chlorine shows the largest deshielding effect. The β effects are nearly the same for all halogens, the least electronegative atom, iodine, as in the case of the alkylhalides,[31] exhibiting the largest effect.

TABLE 6
Substituent Effects of RX^+ Groups in Dialkylhalonium Ions Compared with Parent Alkanes

Carbon position	$\Delta\delta$ for ions from hydrocarbons (ppm)[a]		
	X = Cl	X = Br	X = I
R = Methyl			
Alpha			
CH_3[c]	50.9	39.6	11.5
CH_2[d]	67.5	60.9	31.9
Beta	10.9	10.7	11.3
R = Ethyl			
Alpha			
CH_3[c]	50.0	38.1	10.0
CH_2[d]	67.1	58.4	30.9
Beta	11.5	12.1	12.7

[a]The differences were calculated by subtracting the chemical shifts in a dialkylhalonium ion from the shifts in the corresponding unsubstituted hydrocarbon.
[b]A positive sign indicates deshielding.
[c]Chemical shift difference between CH_4 and CH_3R, where $R = CH_3X^+$ or $CH_3CH_2X^+$.
[d]Chemical shift difference between CH_3CH_3 and CH_3CH_2R, where $R = CH_3X^+$ or $CH_3CH_3X^+$.

Differences in the α-substituent effects between alkyl halides[31] and dialkylhalonium ions are shown in Table 7. α-Substituent effects have been found to correlate roughly with substituent electronegativity. The larger deshielding for RX^+ is consistent with its greater electronegativity compared with X.

The data in Table 7 show that the β-carbon shift in dialkylhalonium ions is shielded from the corresponding carbon in the parent alkyl halides. A similar result was observed for carbon shielding in protonated and parent carboxylic acid esters.[32]

TABLE 7
Differences in Substituent Effects between Haloalkanes and Dialkylhalonium Ions (RX^+CH_3)

Carbon position	Δδ for ions from haloalkenes given in parentheses (ppm)[a]		
	X = Cl	X = Br	X = I
R = Methyl			
Alpha			
CH_3	23.7 (CH_3Cl)	27.3 (CH_3Br)	29.9 (CH_3I)
CH_2	33.5 (EtCl)	38.5 (EtBr)	36.6 (EtI)
Beta	-1.9 (EtCl)	-3.7 (EtBr)	-3.4 (EtI)
Gamma			0.5 (PrI)
R = Ethyl			
Alpha			
CH_3	22.8 (CH_3Cl)	25.8 (CH_3Br)	28.4 (CH_3I)
CH_2	33.1 (EtCl)	36.0 (EtBr)	37.6 (EtI)
Beta	-1.3 (EtCl)	-2.3 (EtBr)	-3.0 (EtI)

[a]Positive values indicate a deshielding in the dialkylhalonium ion.

The cmr shifts of the methyl carbons of dimethylhalonium ions indicate that little charge is introduced into these carbons. The deshielding from their precursors (methyl chloride, bromide, and iodide) is 29.0, 25.8, and 23.0 ppm, respectively. The methyl carbons thus clearly remain sp^3-hybridized. Most of the positive charge on dialkylhalonium ions is therefore located at the halogen centers.

It is interesting to note that carbons attached to chlorine are consistently more deshielded than those attached to bromine and iodine in the homologous dialkylhalonium ions. This is probably due to the fact that the larger halogen (e.g., iodine) can accommodate more charge than the smaller (e.g., chlorine), and also to related magnetic shielding effects (see Section 0.0). The relative stability of dialkylahlonium ions was observed to be in the order $\diagdown\overset{+}{I}\diagup$ > $\diagdown\overset{+}{Br}\diagup$ > $\diagdown\overset{+}{Cl}\diagup$, whereas their relative chemical reactivity had an opposite sequence $\diagdown\overset{+}{Cl}\diagup$ > $\diagdown\overset{+}{Br}\diagup$ > $\diagdown\overset{+}{I}\diagup$ (as observed in their alkylation reactions; see Section 5.4). These observations and the nmr data are thus in good accordance. Stability and reactivity of the ions are thus opposite, and similar, properties.

3.3 RAMAN AND IR SPECTROSCOPIC STUDIES

Olah and DeMember[26b,c] reported Raman and ir spectroscopic studies of dimethylhalonium ions. Raman data for the dimethylhalonium ions studied are summarized in Table 8, along with data observed for isoelectronic model compounds.[33] Dimethylhalonium ions ($CH_3X^+CH_3$) could exist in either a

TABLE 8
Raman Frequencies of Dialkylhalonium Ions, Their Isoelectronic Models, and Methyl Halides[a]

Compound	ν C–X–C deformation	ν_6 [ν_{17}] C–X stretching	C–H stretching
$CH_3\overset{+}{Cl}CH_3(SO_2)$	392 (0.31)	610 (0.36)	2865,2950,2980, 3080
$CH_3\overset{+}{Br}CH_3(SO_2)$	282 (0.46)	544 (0.31)[561 (dp)]	2865(0.70),2950, 2978(0.23),3100
$CH_3\overset{+}{I}CH_3(SO_2)$	?	511 (0.33)	2872,2952,2973
$CH_3CH_2\overset{+}{Cl}CH_2CH_3$ (SO_2)	390 (0.27)	610 (0.35), 545 (0.36), 527 (dp)	2860(0.70),2945 (0.24),2987 (0.65),3060
CH_3SCH_3 (neat)	282 (p)	691 (p)[742(dp)]	2832(p),2910(p), 2966(p),2982 (dp)
CH_3SeCH_3 (neat)	236 (dp)	587 (p)[602(dp)]	2823(p),2919(p), 2996(dp)
CH_3TeCH_3 (neat)	198	526 (dp)	2810,2923(p) 3000(dp)
$CH_3CH_2SCH_2CH_3(SO_2)$	340 (0.54)	693(0.82),656 (0.34),638(0.82)	2946
CH_3Cl (neat) (SO_2)		731.2 706[$\delta(\Delta\nu)$=25cm^{-1}]	1878.8,2966.2, 3041.8,2869, 2969
CH_3Br (neat (SO_2)		611 595[$\delta(\Delta\nu)$=15cm^{-1}]	2861,2972, 3055.8,2869, 2969, ?
CH_3I (neat) (SO_2)		532.8 527[$\delta(\Delta\nu)$=6cm^{-1}]	2861,2969.8, 3060.6,2867, 2968, ?

[a]Spectra reported were recorded at −40°. For detailed experimental conditions see G. A. Olah and A. Commeyras, *J. Am. Chem. Soc.*, **91**, 2929 (1969). Values in parentheses are the depolarization factors taken as an average for three runs and are uncorrected. p, Polarized; dp, depolarized.

linear or bent conformation. On the basis of energetic considerations and the well-established geometry of the isoelectronic dimethyl chalcides, it is expected that dialkylhalonium ions contain sp^3-hybridized halogen atoms, resulting in approximately tetrahedral geometry. Such an arrangement would lead to three skeletal vibrations which are both ir- and Raman-active. Two of the vibrations are totally symmetric (**A**$_1$ in character) and one asymmetric (of the **B**$_1$ type). Thus two polarized and one depolarized Raman-active, and three ir-active modes, are theoretically predicted for the fundamental skeletal vibrations of C_{2v} dimethylhalonium ions.

The observation of three Raman- and three ir-active skeletal vibrations for the dimethylhalonium ion, with two totally symmetric and one asymmetric, in accordance with the selection rules for the $\mathbf{C}_{2v}$ point group, and the correlation with the skeletal frequencies of dimethyl selenide, indicate that the methyl groups maintain preferred conformations with torsional minima in a $\mathbf{C}_{2v}$ arrangement. The Raman spectrum of the dimethylchloronium ion does not parallel that observed for dimethyl sulfide well enough to suggest a significant correlation between their molecular symmetry.

3.4 CHEMICAL REACTIVITY

Dialkylhalonium ions are reactive alkylating reagents. The alkylation of π-donor (aromatic and olefins) and n-donor bases with dialkylhalonium ions was studied.[26c] Alkylation of aromatics with dialkylhalonium ions was found to be not significantly different from conventional Friedel-Crafts alkylations, showing particular similarities in the case of alkylation with alkyl iodides. Alkylation of n-donor bases with dialkylhalonium salts provides a simple synthetic route to a wide variety of onium ions.

3.4.1 Alkylation of Aromatics

The alkylation of aromatic hydrocarbons, such as benzene, toluene, and ethylbenzene, was studied with dimethyl- and diethylhalonium hexafluoroantimonates in SO_2ClF solution.[26c] Data for alkylations are summarized in Table

$$ArH + RX\overset{+}{R}\,SbF_6^- \rightarrow ArR + RX + HSbF_6$$

$$R = CH_3, C_2H_5; \quad ArH = C_6H_6, C_6H_5CH_3, C_6H_5C_2H_5$$

$$X = Cl, Br, I$$

9. Dimethyldihalonium ions give similar methylation results and are quite reactive even at temperatures as low as $-50°$. The dimethyliodonium ion is less reactive and alkylate benzene and toluene in SO_2ClF solution only when allowed to react (if necessary under pressure) at or above $0°$.

TABLE 9

Alkylation of Benzene, Toluene, and Ethylbenzene with Dimethyl- and Diethylhalonium Fluoroantimonates in SO_2ClF[a]

Halonium ion	Aromatic substrate	Temp. (°C)	Time (min)	k_T/k_B	Isomer distribution (%) Ortho	Meta	Para	Ortho/para ratio
$(CH_3)_2Cl^+$	Toluene	25	10		46.6	27.2	26.7	1.75
		0	1		51.8	16.2	32.1	1.61
		-50	5		52.3	15.7	32.0	1.63
		-50	2		58.6	13.0	28.4	2.06
$(CH_3)_2Br^+$	Toluene	-50	5		57.8	9.5	32.7	1.76
		-50	2		59.0	8.6	32.4	1.82
$(CH_3)_2I^+$	Toluene	25	10		46.2	15.6	38.1	1.27
		0	10		53.9	11.8	34.3	1.57
		-20	60		No reaction			
$(CH_3CH_2)_2Cl^+$	Benzene-	-78	1	4.9	33.4	28.1	38.5	0.86
	Toluene	-78	2.5	4.8	31.4	24.6	44.0	0.72
	Ethylbenzene-toluene	-78	2.5	1.10	31.9	19.3	48.8	0.65
$(CH_3CH_2)_2Br^+$	Benzene-	-78	1	4.0	38.7	19.3	42.0	0.92
	Toluene	-78	5	4.5	36.0	18.2	45.8	0.78
	Ethylbenzene-toluene	-78	5	1.2	32.8	14.5	52.8	0.62
					(25.2)	(19.4)	(55.4)	(0.45)
$(CH_3CH_2)_2I^+$		-45	5	4.1	44.0	10.2	45.8	0.96

[a]All data are average of at least three parallel experiments.

The ethylation of toluene by diethylhalonium ions gives ethyltoluenes with ortho/para isomer ratios between 0.60 and 0.96. The ortho/para isomer ratios obtained for the alkylation of toluene in conventional Friedel-Crafts ethylations range from 1.17 to 1.84 (average ca. 1.60). Furthermore, an approximate twofold increase in the k_T/k_B rate ratios for ethylation of toluene (T)-benzene (B) mixtures relative to usual competitive Friedel-Crafts ethylation reactions is also observed. Such differences are considered to be due to the steric ortho effect caused by diethylhalonium ions, and are in accordance with the most probable displacement reaction by the aromatic on the bulky diethylhalonium ions. This can be envisioned to proceed through an S_N2-type transition state.

$$\mathrm{ArH} + \mathrm{CH_3CH_2{-}\overset{+}{X}{-}Et}\ \mathrm{SbF_6^-} \rightleftharpoons \left[\mathrm{HAr{-}{-}{-}{-}CH(CH_3)H{-}{-}{-}{-}X\,Et}\right]^+ \mathrm{SbF_6^-} \rightleftharpoons \left[\mathrm{Ar(H)CH_2CH_3}\right]^+ \mathrm{SbF_6^-} + \mathrm{CH_3CH_2X}$$

$$\longrightarrow \mathrm{ArCH_2CH_3} + \mathrm{HSbF_6}$$

The alkylation data obtained from the reaction of dimethyl- and diethylhalonium ions provide evidence for direct alkylation of aromatics by dialkylhalonium ions. In addition, the data indicate that dialkylhalonium ions are not necessarily involved as active alkylating agents in general Friedel-Crafts systems, although some of the reported anomalous alkylation results, particularly with alkyl iodides, could be attributed to reaction conditions favoring dialkylhalonium ion formation.

3.4.2 Alkylation of Heteroorganic Compounds

Dimethyl- and diethylhalonium ions are particularly effective alkylating agents for heteroorganic compounds.[26c] Table 10 summarizes the pmr data for onium ions obtained by alkylation of ethers, alcohols, water, ketones, aldehydes, and carboxylic acids, as well as nitro, sulfur, and amino compounds with dimethyl-, and diethyldihalonium ions. Alkylation of *O*-, *S*- and *N*- *n*-donor bases by dialkylchloronium and -bromonium ions appears to be practically quantitative, even under mild conditions. Dialkyliodonium ions generally do not react with the same *n*-donor bases in SO_2 or SO_2ClF solution at temperatures between -78 and $0°$ (although they can undergo alkylation under more forceful conditions).

In comparison with Meerwein's trialkyloxonium ions,[34] the advantage of alkylation with dialkylhalonium ions lies in their greater alkylating

TABLE 10

Onium Ions Obtained by Methylation and Ethylation of Heteroatom *n*-Donor Bases with Dimethyl(diethyl)halonium Fluoroantimonates

Heteroorganic substrate	Methylated onium ion	Pmr shifts (δ) CH_3Y	CH_3	CH_2	CH	HY	Aromatic	Ethylated onium ion	Pmr shifts (δ) CH_3-CH_2Y	CH_3-CH_2Y	CH_3Y	CH_3	CH_2	CH	HY	Aromatic
$(CH_3)_2O$	$(CH_3)_3O^+$	4.30						$(CH_3)_2O^+C_2H_5$	4.12	1.27	3.98					
$(C_2H_5)_2O$	$(C_2H_5)_2O^+CH_3$	4.23	1.58	4.62				$(C_2H_5)_3O^+$	4.33	1.41						
CH_3OH	$CH_3O^+(CH_3)H$	3.68				9.42		$CH_3OH(C_2H_5)^+$	4.27	1.39	3.88				9.38	
H_2O	$CH_3OH_2^+$	4.13				9.50		$C_2H_5OH_2^+$	4.86	1.60					9.2	
$(CH_3)_2C{=}O$	$(CH_3)_2C{=}^+OCH_3$	4.80	3.94					$(CH_3)_2C{=}^+OC_2H_5$	5.02	2.05		3.92				
$(C_2H_5)_2C{=}O$	$(C_2H_5)_2C{=}^+OCH_3$	4.63	1.44	3.82				$(C_2H_5)_2C{=}^+OC_2H_5^+$	5.13	1.98		1.41	3.80			
c-C_5H_8O	c-$C_5H_8OCH_3^+$	4.86			3.18, 2.24			c-$C_5H_8OC_2H_5^+$	5.25	1.98						
$(C_6H_5)_2C{=}O$	$(C_6H_5)_2C{=}^+OCH_3$	3.83			9.62			$(C_6H_5)_2C{=}^+OC_2H_5$	4.38	1.08						7.7-8.0
CH_3CHO	$CH_3CH{=}^+OCH_3$	5.80	3.51		9.62			$CH_3CH{=}^+OC_2H_5$	5.21	1.86			3.46	9.60		
C_6H_5CHO	$C_6H_5CH{=}^+OCH_3$	4.12			9.42		7.7-8.9	$C_6H_5CH{=}^+OC_2H_5$	5.58	1.92				9.95		7.7-8.0
$HCOOH$	$HC(OCH_3^+)OH$ syn	4.25			8.76	12.54		$HC(OC_2H_5^+)OH$ syn	4.54	1.92				8.62	13.06	
	anti	4.18			8.70	12.54		anti	5.00	2.06				8.42	13.06	
CH_3COOH	$CH_3C(OCH_3^+)OH$	4.32	2.64			12.84		$CH_3C(OC_2H_5^+)OH$	4.95	1.48		2.75			12.47	
CH_3NO_2	$CH_3NO_2^+CH_3$	4.67	5.04					$CH_3NO_2^+C_2H_5$	5.15	1.70		5.02				
$CH_3CH_2CH_2NO_2$	$CH_2CH_2CH_2NO_2^+CH_3$	4.67	0.98	5.23, 2.18				$CH_3CH_2CH_2NO_2^+C_2H_5$	5.14	1.70		0.99	5.20, 2.16			
$C_6H_5NO_2$	$C_6H_5NO_2^+CH_3$	4.65					7.6-8.2	$C_6H_5NO_2^+C_2H_5$	5.12	1.69						7.6-8.2
$((CH_3)_2CH)_2S$	$((CH_3)CH)_2S^+CH_3$	2.52	1.38		3.53			$[(CH_3)CH]_2S^+C_2H_5$	3.03	1.45		1.33		3.57		
$(C_2H_5)_2S$	$(C_2H_5)_2S^+CH_5$	2.76	1.38	3.18				$(C_2H_5)_3S^+$	3.26	1.40						
t-BuSH	*t*-BuS^+HCH_3	2.99	1.62			6.16		*t*-$BuS^+C_2H_5$	3.58	1.54		1.63			6.08	
$(C_2H_5)_3N$	$(C_2H_5)_3N^+CH_3$	2.97	1.23	3.21				$(C_2H_5)_4N^+$	4.81	1.05						

ability (in case of chloronium and bromonium ions) and in the possibility of increased selectivity when the halogen center is being changed (iodonium ions are generally the most stable but the least reactive).

3.5 INTERMOLECULAR EXCHANGE REACTIONS OF DIALKYLHALONIUM IONS WITH ALKYL HALIDES AND ALKYLCARBENIUM IONS

We have so far discussed the alkylation of a variety of *n*-donor bases, other than halogen donors, by dimethyl- and diethylhalonium ions. If the *n*-donor base is an excess of alkyl halide, intermolecular exchange reactions via transalkylation take place.[27] The intermolecular exchange reaction of dialkylhalonium ions with alkyl halides and alkylcarbenium ions[35] was studied in detail.

$$RX^{+}R + RX \rightleftharpoons RX + RX^{+}R$$

Dimethyl- and diethylhalonium ions, prepared from methyl and ethyl halides with SbF_5 in SO_2 solution, do not undergo intermolecular exchange reactions with excess methyl and ethyl halides. Similarly, dimethyl- and diethylhalonium ions, prepared by methylation of methyl halides and ethylation of ethyl halides with methyl and ethyl fluoroantimonates,[36] do not undergo intermolecular exchange reactions with an excess of methyl and ethyl fluorantimonate, respectively.

$$RX^{+}R\ SbF_5X^{-} + {}^{*}R^{+}SbF_6^{-} \rightleftharpoons RX^{+}R^{*}\ SbF_5X^{-} + R^{+}SbF_6^{-}$$

$$R = R^{*} = CH_3,\ C_2H_5;\ X = Cl,\ Br,\ I$$

$$RX^{+}R\ SbF_5X^{-} + R^{*}X \rightleftharpoons RX^{+}R^{*}\ SbF_5X^{-} + RX$$

$$R = R^{*} = CH_3,\ C_2H_5;\ X = Cl,\ Br,\ I$$

Diisopropylchloronium ion undergoes intermolecular exchange with both isopropyl cation and isopropyl chloride. The exchange reactions depend on the experimental conditions and can be divided into three types: (1) exchange with isopropyl chloride when 2 equivalents or more of isopropyl chloride are treated with SbF_5 in SO_2 solution, (2) exchange with isopropyl cation when

$$[(CH_3)_2CH]_2Cl^{+} + CH_3CHClCH_3 \rightleftharpoons CH_3CHClCH_3 + [(CH_3)_2CH]_2Cl^{+} \quad (1)$$

$$[(CH_3)_2CH]_2Cl^{+} + CH_3CH^{+}CH_3 \rightleftharpoons CH_3CH^{+}CH_3 + [(CH_3)_2CH]_2Cl^{+} \quad (2)$$

isopropyl chloride is treated with 2-4 equivalents of SbF_5 (isorpopyl cation is formed when 4 equivalents or more of SbF_5 are used), and (3) when 0.5-2.0 equivalents of isopropyl chloride are treated with SbF_5 in SO_2 solution; three different intermolecular exchange reactions, proceeding through carbonium ion transition states, occur in addition to the exchange reactions (1) and (2).

$$\begin{array}{c} CH_3\overset{+}{C}ClCH_3 \\ + \\ CH_3CH_2CH_3 \end{array} \overset{b}{\rightleftharpoons} \left[\begin{array}{c} \text{H} \quad \text{H} \quad CH_3 \\ CH_3{-}C\overset{a}{\cdots}\overset{c}{\vdots}\overset{b}{\cdots}C{-}CH_3 \\ CH_3 \qquad\qquad Cl \end{array} \right]^+ \overset{a}{\rightleftharpoons} (CH_3)_2\overset{+}{C}H + CH_3CHClCH_3 \rightleftharpoons [(CH_3)_2CH]_2Cl^+ \quad (3)$$

$$\overset{c}{\rightleftharpoons} CH_3{-}\underset{CH_3}{\overset{H}{C}}{-}\underset{CH_3}{\overset{Cl}{C}}{-}CH_3 + H^+ \rightleftharpoons CH_3{-}\underset{CH_3}{\overset{H}{C}}{-}\underset{CH_3}{\overset{+}{C}}{-}CH_3 + HCl$$

The last type [equation (3)] of intermolecular exchange reaction is particularly significant, because it shows the competing σ- versus n-donor ability toward an electrophile. The isopropyl cation alkylates the n-donor chlorine atom of isopropyl chloride to give the diisopropylchloronium ion. At the same time it is also capable of alkylating the σ-donor C—H single bond[35] of isopropyl chloride, forming a carbonium ion-type transition state. There are three possible types of cleavage (*a*, *b*, and *c*). The carbonium ion can return to isopropyl chloride and isopropyl cation via *a*. Protolytic cleavage via *b* could give dimethylchlorocarbenium ion and propane. This has been experimentally proved by mixing an appropriate amount of dimethylchlorocarbenium ion and propane and obtaining the same reaction mixture, as shown by pmr spectroscopy. Deprotonation could generate the alkylation product 2,3-dimethyl-2-chlorobutane, which is further ionized to *tert*-hexyl cation. The formation of *tert*-hexyl cation is an irreversible process.

Diisopropylbromonium and diisopropyliodonium ions also exchange with excess isopropyl bromide and iodide, respectively. Such exchange reactions

$$[(CH_3)_2CH]_2X^+ + CH_3CHXCH_3 \rightleftharpoons CH_3CHXCH_3 + [(CH_3)_2CH]_2X^+$$
$$X = Br, I$$

have been studied by temperature-dependent pmr spectra. Diisopropylbromonium ions also exchange with excess isopropyl cations, similarly to the exchange of diisopropylchloronium ions. Diisopropyliodonium ion, however, does not exchange with isopropyl cation even when 4 equivalents or more of

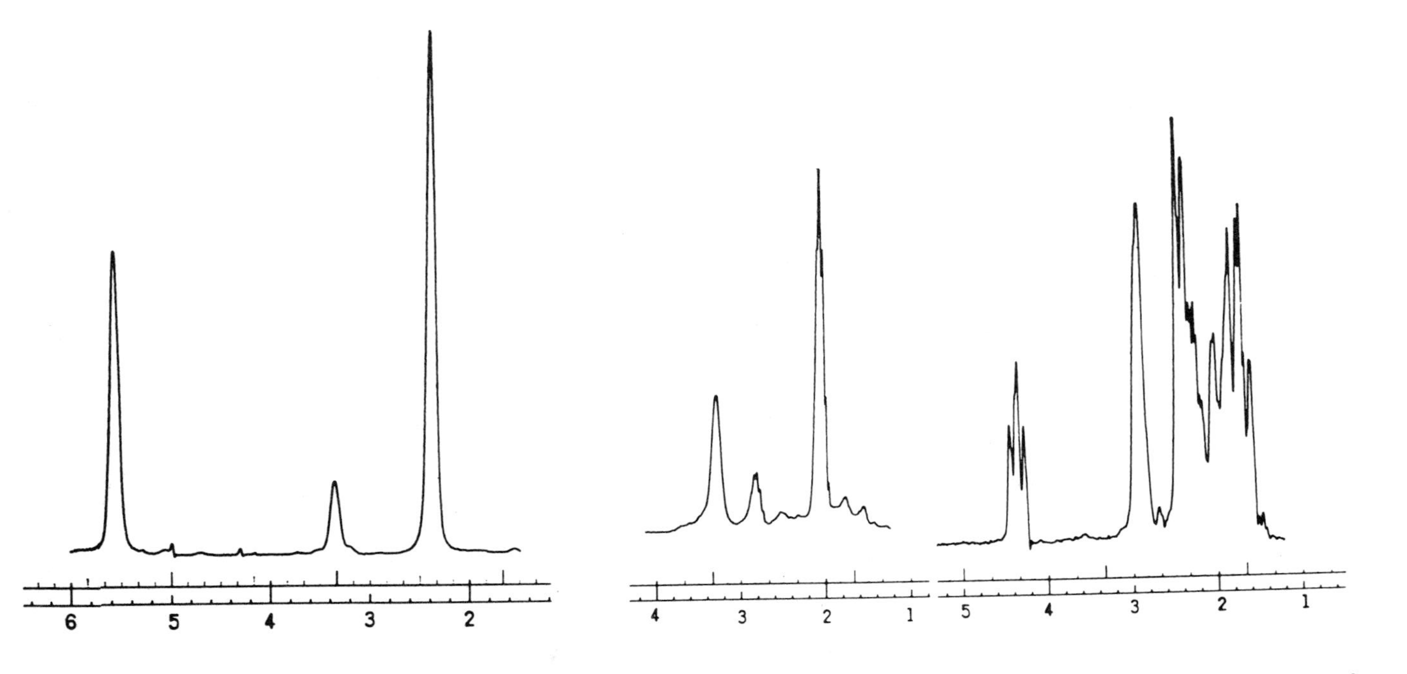

Figure 1. Pmr spectra of norbornyl cation (left), exchanging dinorbornylchloronium ion (center), and 2-*exo*-norbornyl chloride (right) at $-78°$.

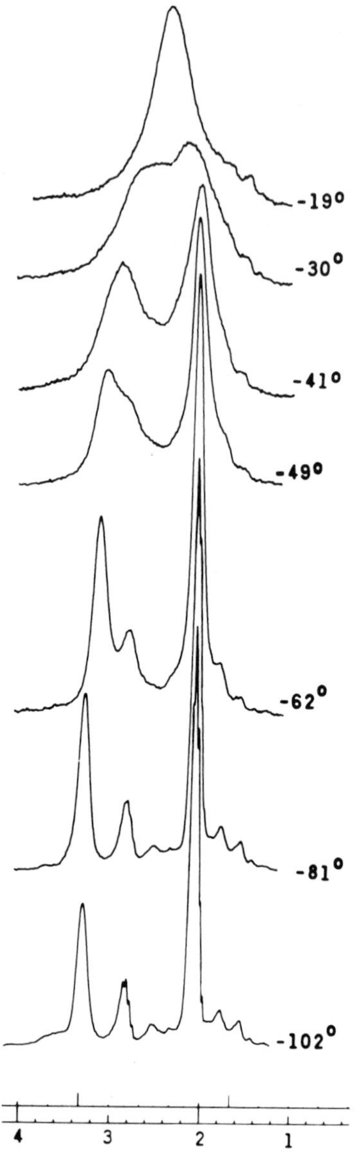

Figure 2. Temperature-dependent pmr spectra of the dinorbornylchloronium ion.

SbF_5 is treated with isopropyl iodide. Only diisopropyliodonium ion was observed under these experimental conditions. Isopropyl cation could not be generated from isopropyl iodide under any stable ion conditions utilized. These data reflect the unusual stability of the diisopropyliodonium ion.

The intermolecular exchange reaction of di-*tert*-butylhalonium ions and di-*tert*-amylhalonium ions with their corresponding halides or/and alkylcarbenium ions was also studied. These exchange reactions have been demonstrated by the concentration-dependent pmr spectra of the systems.

$$R(CH_3)_2CX^+C(CH_3)_2R + R(CH_3)_2C^+ \rightleftharpoons R(CH_3)_2C^+ + R(CH_3)_2CX^+C(CH_3)_2R$$

$$R(CH_3)_2CX^+C(CH_3)_2R + R(CH_3)_2CX \rightleftharpoons R(CH_3)_2CX + R(CH_3)_2CX^+C(CH_3)_2R$$

$$R = CH_3;\ X = Br,\ Cl$$

$$R = C_2H_5;\ X = Br,\ Cl$$

The intermolecular exchange reactions of dinorbornylhalonium ions are interesting. These ions exchange with either excess norbornyl cation or excess 2-*exo*-norbornyl halide, depending on the experimental conditions. The pmr spectra of all the systems are similar to that of the norbornyl cation,[37] regardless of which component is in excess, and are temperature-dependent (Figures 1 and 2). The pmr absorptions are deshielded when more SbF_5 is added and are shielded when more 2-*exo*-norbornyl halide is added. The

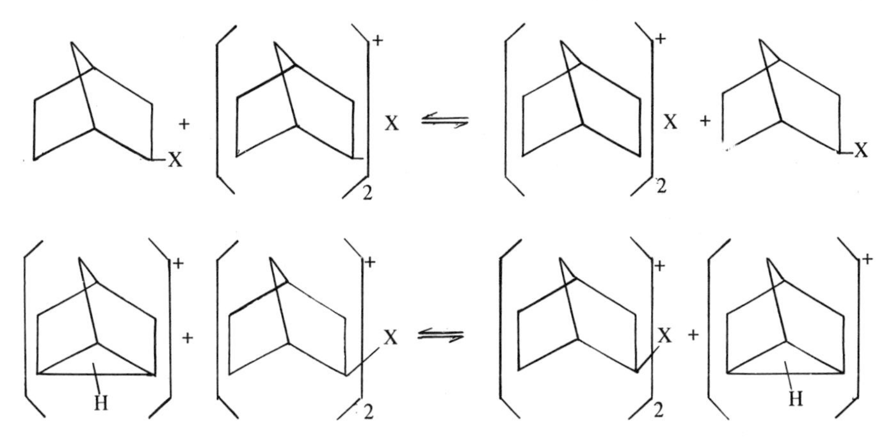

temperature-dependent nature of the systems indicates that 6,1,2-hydrogen shifts, as well as the Wagner-Meerwein rearrangement, are still taking place in the involved unassociated norbornyl cations at the temperatures studied ($> -100°$), in accordance with studies of the norbornyl cation itself.

The 1-adamantyl cation prepared from adamantyl halides with SbF_5-SO_2SO_2ClF) solution[38] displays three pmr absorptions at δ 5.40, 4.52, and 2.67. Diadamantylhalonium ions were formed as yellowish-orange precipitates when equimolar amounts of adamantyl halides were added to a solution of adamantyl cations in SbF_5-SO_2 (SO_2ClF). The precipitates were redissolved when methylene chloride was added as a cosolvent. The pmr spectrum of each of the resulting solution showed three absorptions similar to those of the adamantyl cation but slightly shielded. The shielding effect is proportional to the amount of adamantyl chloride or bromide added to the solution of the adamantyl cation. For example, Figure 3 (left) shows the change in proton chemical shifts when adamantyl chloride was added to a solution of the 1-adamantyl cation. Further addition of

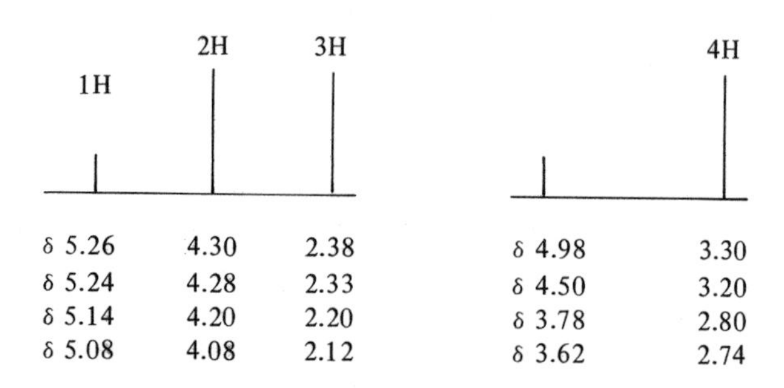

Figure 3. Pmr data of diadamantylchloronium ion exchanging with 1-adamantyl chloride and/or 1-adamantyl cation.

adamantyl chloride to the above solution caused the collapse of the two methylene absorptions into a singlet. The new singlet and the methine absorption both became shielded when the concentration of adamantyl chloride was increased (Figure 3, right). The ratio of the two proton absorptions is 4:1. The pmr spectrum is temperature-independent from -80 to $-10°$. These data again suggest intermolecular exchange of diadamantylchloronium ion with adamantyl chloride and/or adamantyl cation. In this case the 1-adamantyl cation can also undergoes intermolecular hydrogen exchange with a tertiary C—H bond of 1-haloadamantanes (the σ-donor site competing with the n-donor halogen atoms).

$$\mathrm{AdX^{+}Ad} + \mathrm{Ad^{+}} \rightleftharpoons \mathrm{Ad^{+}} + \mathrm{AdX^{+}Ad}$$
$$\mathrm{AdX^{+}Ad} + \mathrm{AdX} \rightleftharpoons \mathrm{AdX} + \mathrm{AdX^{+}Ad}$$
$$\mathrm{Ad^{+}} + \mathrm{AdX} \rightleftharpoons \mathrm{AdH} + \mathrm{XAd^{+}}$$

Ad = adamantyl; X = Br, Cl

3.5.1 Determination of Relative Carbocation Stabilities through Competitive Exchange

The intermolecular exchange reaction of alkyl halides with alkylcarbenium ions through dialkylhalonium ions can be advantageously used to determine directly, by competitive comparison, the relative stabilities of the ions.[39]

Tertiary carbenium ions are uniformally accepted to be considerably more stable than secondary ions.[35] Should a secondary ion exhibit extraordinary stability, a specific reason, such as nonclassical σ-delocalization, would be required to explain its behavior. The much-studied 2-norbornyl cation[37] indeed possesses such unusual stability as compared with other secondary or even tertiary ions. This conclusion is based on the direct observation of the preferred equilibria formed between a series of secondary and tertiary carbenium ions and 2-norbornyl chloride and bromide, respectively, in SO_2ClF solution at −70° (as determined by pmr spectroscopic study of the ions present).

X + R^+ ⇋ [X^+—R] ⇋ + + R—X

R^+ =$(CH_3)_3C^+$,$(CH_3)_2CH^+$, cyclopentyl, 1-methyl-1-cyclopentyl, and 1-adamantyl cations
X = Cl and Br

The equilibria shown in Table 11 were also determined from the other direction, that is, by the addition of solutions of alkyl or cycloalkyl chlorides and bromides in SO_2ClF to a solution of the norbornyl cation (also in a SO_2ClF solution without excess SbF_5) at −70° and followed by pmr. The interchange reaction is considered to involve formation of the corresponding alkylnorbornylhalonium ions which then disproportionate in the direction of the thermodynamically more stable 2-norbornyl or alkyl cations.

When the secondary 2-norbornyl cation was allowed to compete in SbF_5-SO_2ClF solution with tertiary carbenium ions, such as *tert*-butyl, methylcyclopentyl, and 1-adamantyl cations, the equilibria formed show the former to exist as the predominant ion, or at least comparable in concentration to the latter ions, indicating that the 2-norbornyl cation is more or at least as stable as the compared tertiary carbocations. When the 2-norbornyl cation was allowed to compete with secondary carbocations, such as isopropyl and cyclopentyl cations, the former appear as the sole detectable ion in the solution (see Table 11).

TABLE 11
Equilibrium Concentration of Competing Carbocations at −70° in SO_2ClF Solution

R^+	R^+ + *exo*-2-norbornyl bromide ⇌	R bromide + 2-norbornyl cation
tert-Butyl	30%	70%
1-Methyl-1-cyclopentyl	20%	80%
1-Adamantyl	45%	55%
Isopropyl	(<2%)	>98%
Cyclopentyl	(<2%)	>98%

The 2-norbornyl cation thus possesses not only comparable or greater stability than tertiary alkyl or cycloalkyl cations, but also greater stability than the degenerate, rapidly equilibrating secondary cyclopentyl cation.[35] Rapid equilibration thus provides no substantial stabilization in itself. When competing with the *tert*-butyl cation, the cyclopentyl cation was also found to be considerably less stable.

The results are in good accord with recent gas-phase ion stability measurements by Beauchamp[40] using competitive ICR techniques (with which of course no structure for the ions involved can be assured). Both in solution and in the gas phase the norbornyl ion thus shows exceptional stability when compared to usual secondary carbocations, and is similar or more stable than tertiary ions. Only the σ-delocalized nonclassical structure of the ion can account for these findings, in complete agreement with all other structural and chemical studies.

Direct competitive comparison of carbocation stability through exchange equilibria of alkyl or cycloalkyl halides and alkyl- or cycloalkylcarbenium ions is also applicable to other systems and should gain extended use. As the exchange proceeds through dialkyl- or cycloalkylhalonium ions utilizing the nonbonded electron pairs of halogens, steric hindrance should not much affect the reactions, which could be a major consideration in exchange reactions of carbocations with the corresponding hydrocarbons (σ donors).

3.6 ALKYLATION AND POLYMERIZATION OF ALKENES

Dialkylhalonium ions, as discussed, are effective alkylating agents for aromatics and various *n*-donor bases. A study of the reaction of alkenes such as 1-butene, 2-butene, and 2-methylpropene (isobutylene) with dimethylbromonium fluoroantimonate in SO_2 solution has also been carried out.[26a,41] Ready alkylation takes place but under the reaction conditions, the initially formed alkylcarbenium ions are not sufficiently stable to be observed. In the presence of excess alkenes, the reaction proceeds rapidly, yielding polymers.[41]

$$(CH_3)_2C{=}CH_2 + R_2X^+ \rightarrow (CH_3)_2\overset{+}{C}{-}CH_2R + RX \xrightarrow{(CH_3)_2C=CH_2} \text{polymer}$$

Dialkylhalonium ions thus are effective cationic polymerization initiators. The possible involvement of dialkylhalonium ions in Friedel-Crafts polymerization of alkenes with systems such as $AlCl_3$-CH_3Cl and $AlBr_3$-CH_3Br must therefore also be considered.

The aluminum halide-catalyzed polymerization of isobutylene (the most extensively studied cationic polymerization) is frequently carried out in methyl chloride solution. The same reaction in methyl bromide is much slower, and methyl iodide practically stops the polymerization. Plesch[42] used the slowness of the reaction in methyl bromide to develop his concept that the catalytic activity of aluminum bromide in the system is due to the $AlBr_2^+$ ion. The real reason, however, probably is formation of dialkylhalonium ions not considered before, with exception of some of Kennedy's[43] recent work in a somewhat different context. It was indeed shown that *n*-donor alkyl halides can compete with π-donor alkenes for alkylcarbenium ions and slow down polymerization parallel with the stability of the corresponding dimethylhalonium ions, and follow the order of decreasing deactivating trend: $CH_3I > CH_3Br > CH_3Cl$.

3.7 ROLE OF DIALKYLHALONIUM IONS IN FRIEDEL–CRAFTS REACTIONS

The observation of dialkylhalonium ion formation in alkyl halide-antimony pentafluoride systems raises the question of their involvement in Friedel-Crafts reactions (alkylations and polymerizations). Alkyl halide-Lewis acid halide complexes (as well as the related alkylcarbenium ions) can obviously alkylate not only aromatic and aliphatic hydrocarbons but also excess alkyl halides (other than fluorides) to form dialkylhalonium ions. The 1:1 or 1:2 alkyl halide-Lewis acid complexes therefore could be expected to contain dialkylhalonium ion complexes.

DeHaan and Brown[44] showed, based on kinetic evidence, that in methyl bromide-gallium bromide (and related chloride systems) the dimethylbromonium (or -chloronium) ion is involved.

$$\overset{\delta^+}{RCl} \rightarrow \overset{\delta^-}{AlCl_3} \qquad\qquad \overset{\delta^+}{RCl} \rightarrow \overset{\delta^-}{Al_2Cl_6}$$

$$1:1 \qquad\qquad 1:2$$

$$RCl^+\text{-}R\ \ Cl \rightarrow AlCl_3^- \qquad\qquad RCl^+\text{-}R\ \ Cl \rightarrow Al_2Cl_6^-$$

$$2:1 \qquad\qquad 2:2$$

The preparation and study of a series of dialkylhalonium complexes allowed a direct comparison with common Friedel-Crafts alkylation systems, such as methyl halide-aluminum halide complexes, in order to determine directly the possible contribution of dimethylhalonium ions.

$AlBr_3$ dissolves in CH_3Br and gives a clear solution which is stable at room temperature. Walker,[45] from vapor pressure measurements, found the composition of the liquid phase to be $CH_3Br\text{-}AlBr_3/CH_3Br\text{-}Al_2Br_6 = 0.678:0.288 = 0.033$ (at 23.8°). By the same technique Brown and Wallace[46] established the 1:1 complex at −78°. Ir spectra of this complex were recorded by Perkampus and Baumgarten[47] at −196°, and by Kinsella and Coward[48] at room temperature. Raman investigations were carried out by Rice and Bold.[49] However, no clear conclusions were reached from these studies.

A comparative study with nmr spectroscopy of the 1:1 CH_3-Br-$AlBr_3$ complex and the related 1:1 AlI_3-CH_3I complex (the $AlCl_3$-CH_3Cl complex is less suitable, because of the lower boiling point of methyl chloride and the difficulties in handling the 1:1 complex at atmospheric pressure) showed in the 1:1 CH_3Br-$AlBr_3$ complex (neat) a singlet pmr absorption at δ 4.0 (in SO_2ClF solution at -40°, δ 3.40). The corresponding 1:1 CH_3I-AlI_3 complex showed a singlet absorption at δ 3.65. The chemical shifts of both complexes closely resemble those observed for the dimethylbromonium and -iodonium ions, respectively, and are different from the well-defined $CH_3F \rightarrow SbF_5$ donor-acceptor complex[36] (δ 5.56 in SO_2), in which case no dimethylfluoronium ion formation is involved. These data seem to indicate that 1:1 CH_3Br-$AlBr_3$ or CH_3I-AlI_3 complexes contain at least a substantial contribution from the corresponding dimethylbromonium or -iodonium ions in the equilibrating systems.

$$2CH_3Br + Al_2Br_6 \rightleftharpoons CH_3Br \rightarrow Al_2Br_6 + CH_3Br \rightleftharpoons CH_3\overset{+}{Br}CH_3Al_2Br_7^-$$

$$\searrow\!\!\nwarrow \qquad \nearrow\!\!\swarrow$$

$$2(CH_3Br \rightarrow AlBr_3)$$

The chemical behavior of the methyl halide-aluminum halide complex is indicative of the equilibrium system. When treated with typical *n*-donors such as nitromethane, acetone, methyl benzoate, or propionaldehyde, which have been *n*-alkylated with dimethylhalonium ions, the methyl halide complexes in sharp contrast give no alkylation products but form the corresponding $AlBr_3$ or AlI_3 donor-acceptor complexes. This is in accordance with the transfer of aluminum halide from the methyl halide donor-acceptor complexes to the stronger *n*-donor bases, thus shifting the equilibrium in this direction. At the same time, the methyl halide-aluminum halide complexes readily alkylate aromatics or olefins. Hexamethylbenzene, for example, forms heptamethylbenzenium bromoaluminate.[50]

$$C_6(CH_3)_6 + 2(CH_3Br\text{-}AlBr_3) \rightleftharpoons C_6(CH_3)_7^+$$

$$\downarrow\uparrow$$

$$CH_3BrCH_3Al_2Br_7^-$$

δ 3.82

δ 2.60

δ 3.65

δ 3.40

Preparation as well as spectroscopic study and investigation of the alkylating ability of dialkylhalonium ions established their structure and relative stability, as well as their chemical reactivity. In general, the relative stability of dialkylhalonium ions follows the order $RI^+R > RBr^+R > RCl^+R$, indicating that a larger halogen atoms is more capable of accommodating charge. Alkylating ability, however, is generally inversely proportional to the stability of the ion.

Dialkylhalonium ions are also formed under regular Friedel-Crafts conditions as rapidly exchanging systems with excess alkyl halides, according to the equilibria

$$R{-}X^{+}{-}R[M_xY_y]^{-} \rightleftharpoons RX + RX \rightarrow M_xY_{y-1} \qquad \begin{array}{l} M = \text{Al, Ga, etc.} \\ Y = \text{I, Br, Cl} \end{array}$$

Whereas this observation helps to explain the generally very complex kinetic order of Friedel-Crafts alkylation systems, dialkylhalonium ions are not necessarily the de facto reactive alkylating agents. It is clear, however, that even in competition with reactive π-donor systems (olefins, aromatics) an excess of alkyl halide (over the alkyl halide alkylating agent) must be considered an n-donor system capable of forming dialkylhalonium ions. This is not the case, however, for alkyl fluorides. What is interesting, as shown in the case of CH_3F (excess) with SbF_5, is that alkylation of the C–H bond (σ donor) takes place (giving ethyl and subsequently higher alkyl products), as the corresponding dimethylfluoronium ion ($CH_3F^+CH_3$) seems incapable of forming in solution density.

3.8 DIALKYL AKYLENEDIHALONIUM IONS

Alkylation of alkylene dihalides with methyl and ethyl fluoroantimonate (CH_3F-SbF_5-SO_2 and C_2H_5F-SbF_5-SO_2) gives monoalkylated halonium ions and/or dialkylated dihalonium ions, depending on the reaction conditions.[13] Ions prepared are summarized in Table 12. Iodine shows an unusual ability to stabilize positive charge, as demonstrated by the formation of dialkyl alkylenediiodonium ions $RI^+(CH_2)_nI^+R$, n = 1 to 6, R = CH_3, C_2H_5]. In the extreme case (n = 1) the two positive iodonium cation sites are separated by only a single methylene group. However, dialkyl alkylenedibromonium ions were formed only when the two positive bromines were separated by three methylene groups. In the case of four methylene groups, rearrangement to the more stable five-membered-ring tetramethylenebromonium ion took place. Dialkyl alkylenedichloronium ions have not yet been directly observed. Consequently, the ease of formation of dialkyl alkylenedihalonium ions is similar to that of dialkylhalonium ions, that is, $>\overset{+}{I} > >\overset{+}{Br} > >\overset{+}{Cl}$.

The pmr data of mono- and dihalonium ions, formed by alkylation of the corresponding dihaloalkanes under various conditions, are summarized in Table 13. Representative spectra are shown in Figure 4. It is interesting to note that the methyl protons ($-X^+CH_3$) and the methylene protons of ethyl groups ($-X^+CH_2-C$) in all these halonium ions show a deshielding order of $>\overset{+}{Cl} > >\overset{+}{Br} > >\overset{+}{I}$, while the methyl protons of the same ethyl group show an opposite trend $>\overset{+}{I} > >\overset{+}{Br} > >\overset{+}{Cl}$. The structure of dialkyl alkylenedihalonium ions was also studied by carbon-13 nmr spectroscopy (Table 14).

TABLE 12

Alkylation of Dihaloalkanes with Methyl and Ethyl Fluoroantimonate

Dihaloalkane	Alkylating reagent, $RF\text{-}SbF_5\text{-}SO_2$	Halonium ion formation	Remarks
CH_2Cl_2	CH_3 (excess)	$CH_3Cl^+CH_2Cl$	
	C_2H_5 (excess)	$CH_3CH_2Cl^+CH_2Cl$	
CH_2Br_2	CH_3 (excess)	$CH_3Br^+CH_2Br$	
	C_2H_5 (excess)	$CH_3CH_2Br^+CH_2Br$	
CH_2I_2	CH_3 (1 mole)	$CH_3I^+CH_2I$	
	C_2H_5 (1 mole)	$CH_3CH_2I^+CH_2I$	
	CH_3 (excess)	$CH_3I^+CH_2I^+CH_3$	
	C_2H_5 (excess)	$CH_3CH_2I^+CH_2I^+CH_2CH_3$	
$ClCH_2CH_2Cl$	CH_3 (excess)	$CH_3Cl^+CH_3$ and unidentified products	
$BrCH_2CH_2Br$	CH_3 (excess)	$CH_3Br^+CH_3$ and unidentified products	
ICH_2CH_2I	CH_3 (1 mole)	$CH_3I^+CH_2CH_2I$	Forms as a precipitate; structure not yet confirmed
	C_2H_5 (1 mole)	$CH_3CH_2I^+CH_2CH_2I$	Forms as a precipitate; structure not yet confirmed
	CH_3 (excess)	$CH_3I^+CH_2CH_2I^+CH_3$	
	C_2H_5 (excess)	$CH_3CH_2I^+CH_2CH_2I^+CH_2CH_3$	
CH_3CHCl_2	CH_3 (excess)	$CH_3Cl^+CH(Cl)CH_3$	Halogen exchanges to CH_3CHF_2 + $(CH_3)_2\overset{+}{C}$ at −30°
CH_3CHBr_2	CH_3 (excess)	$CH_3CHF_2 + CH_3Br^+CH_3$	Halogen exchanges to CH_3CHF_2 + $(CH_3)_2Br$ at −78°
$ClCH_2CH_2CH_2Cl$	CH_3 (excess)	$CH_3Cl^+CH_2CH_2CH_2Cl$	
	C_2H_5 (excess)	$CH_3CH_2Cl^+CH_2CH_2CH_2Cl$	
$BrCH_2CH_2CH_2Cl$	CH_3 (excess)	$CH_3Br^+CH_2CH_2CH_2Cl$	
	C_2H_5 (excess)	$CH_3CH_2Br^+CH_2CH_2CH_2Cl$	

$Br(CH_2)_3Br$	CH_3 (1 mole)	$CH_3Br^+(CH_2)_3Br$	
	C_2H_5 (1 mole)	$CH_3CH_2Br^+(CH_2)_3Br$	
	CH_3 (excess)	$(CH_3Br^+CH_2)_2CH_2$	
	C_2H_5 (excess)	$(CH_3CH_2Br^+CH_2)_2CH_2$	
$I(CH_2)_3I$	CH_3 (1 mole)	$CH_3I^+(CH_2)_3I$	
	C_2H_5 (1 mole)	$CH_3CH_2I^+(CH_2)_3I$	
	CH_3 (excess)	$(CH_3I^+CH_2)_2CH_2$	
	C_2H_5 (excess)	$(CH_3CH_2I^+CH_2)_2CH_2$	
$CH_3CHBrCH_2Br$	CH_3 (1 mole)	$CH_3CH(Br^+CH_3)CH_2Br \rightleftharpoons CH_3CHBrCH_2Br^+CH_3$	
	CH_3 (excess)	$CH_3CH(Br^+CH_3)CH_2Br^+CH_3$	
$CH_3CHClCH_2Cl$	CH_3 (excess)	$CH_3CH(Cl^+CH_3)CH_2Cl \rightleftharpoons CH_3CHClCH_2Cl^+CH_3$	Rearranges to $(CH_3)_2\overset{+}{Cl}$ and $CH_3\overset{+}{C}HCH_2Cl \rightleftharpoons CH_3CH{-}CH_2$ (bridged $\overset{+}{Cl}$) $\rightleftharpoons$ $CH_3CH_2\overset{+}{C}HCl$ at -10°
$ClCH_2CHClCH_2CH_2CH_3$ or $ClCH_2CH_2CHClCH_3$ or $ClCH_2CH_2CH_2CH_2Cl$	CH_3 (excess)	[tetrahydrofuran] + $(CH_3)_2\overset{+}{Cl}$	Initially formed halonium ions were not observed at -78°
$BrCH_2CHBrCH_2CH_3$ or $BrCH_2CH_2CHBrCH_3$ or $BrCH_2CH_2CH_2CH_2Br$	CH_3 (excess)	[tetrahydrofuran] + $(CH_3)_2\overset{+}{Br}$	Initially formed halonium ions were not observed at -78°
$ICH_2CH_2CH_2CH_2I$	CH_3 (excess)	$CH_3I^+CH_2CH_2CH_2CH_2I^+CH_3$	Rearranges to [tetrahydrofuran] and $(CH_3)_2\overset{+}{I}$ at -40°
$CH_3CHBrCHBrCH_3$ (*meso* and *dl*)	CH_3 (excess)	$(CH_3)_2\overset{+}{Br}$ + [2,3-epoxybutane] + [2,3-epoxybutane]	Initially formed halonium ions were not observed at -78°
$Br(CH_2)_5Br$	CH_3 (excess)	$CH_3Br^+(CH_2)_5Br^+CH_3$	
$I(CH_2)_5I$	CH_3 (excess)	$CH_3I^+(CH_2)_5I^+CH_3$	
$I(CH_2)_6I$	CH_3 (excess)	$CH_3I^+(CH_2)_6I^+CH_3$	

TABLE 13

Pmr Parameters of Mono- and Dialkylated Dihaloalkanes[a]

Ion	δ_{CH}	δ_{CH_2}	δ_{CH_2X}	$\delta_{CH_2\overset{+}{X}}$	$\delta_{\overset{+}{X}CH_2}$	δ_{CH_3}	$\delta_{\overset{+}{X}CH_3}$
$ClCH_2\overset{+}{Cl}CH_3$				5.56 (s)			4.51 (s)
$ClCH_2\overset{+}{Cl}CH_2CH_3$				5.50 (s)	5.52 (q), $J = 7$	2.01 (t), $J = 7$	
$BrCH_2\overset{+}{Br}CH_3$				6.33 (s)			4.30 (s)
$BrCH_2\overset{+}{Br}CH_2CH_3$				6.32 (s)	5.46 (q), $J = 7$	2.22 (t), $J = 7$	
$ICH_2\overset{+}{I}CH_3$				5.17 (s)			3.58 (s)
$ICH_2\overset{+}{I}CH_2CH_3$				5.40 (s)	4.90 (q), $J = 7$	2.40 (t), $J = 7$	
$(CH_3\overset{+}{I})_2CH_2$				5.80 (s)			4.10 (s)
$(CH_3CH_2\overset{+}{I})_2CH_2$				5.88 (s)	5.33 (q), $J = 7$	2.44 (t), $J = 7$	
$(CH_3\overset{+}{I}CH_2)_2$				5.06 (s)			4.00 (s)
$(CH_3CH_2\overset{+}{I}CH_2)_2$				4.90 (s)	4.93 (q), $J = 7.5$	2.20 (t), $J = 7.5$	
$CH_3CH(Cl)\overset{+}{Cl}CH_3$	6.23 (q), $J = 6$					2.17 (d), $J = 6$	4.50 (s)
$Cl(CH_2)_3\overset{+}{Cl}CH_3$		2.90 (qu), $J = 5$	4.00 (t), $J = 5$	5.47 (*t*), $J = 5$			4.52 (s)
$Cl(CH_2)_3\overset{+}{Cl}CH_2CH_3$		2.74 (m)	3.91 (t), $J = 6$	5.3 (m)[b]	5.3 (m)[b]	2.00 (t), $J = 7$	
$Cl(CH_2)_3\overset{+}{Br}CH_3$		2.70 (qu), $J = 5$	3.84 (t), $J = 5$	5.10 (t), $J = 5.5$			4.08 (s)
$Cl(CH_2)_3\overset{+}{Br}CH_2CH_3$		2.83 (m)	3.94 (t), $J = 6$	5.2 (m)[b]	5.2 (m)[b]	2.20 (t), $J = 6.5$	

$Br(CH_2)_3\overset{+}{Br}CH_3$	2.82 (qu), $J = 7$	3.70 (t), $J = 7$	5.20 (t), $J = 6$			4.10 (s)
$Br(CH_2)_3\overset{+}{Br}CH_2CH_3$	2.93 (m)	3.80 (t), $J = 7$	5.1 (m)[b]	5.1 (m)[b]	2.17 (t), $J = 7$	
$I(CH_2)_3\overset{+}{I}CH_3$	2.70 (qu), $J = 7$	3.58 (t), $J = 7$	4.57 (t), $J = 7$			3.58 (s)
$I(CH_2)_3\overset{+}{I}CH_2CH_3$	2.90 (m)	3.92 (t), $J = 7$	4.7 (m)[b]	4.7 (m)[b]	2.28 (t), $J = 7$	
$CH_2(CH_2\overset{+}{Br}CH_3)_2$	3.24 (qu), $J = 7$		5.03 (t), $J = 7$	5.03 (t), $J = 7$		4.26 (s)
$CH_2(CH_2\overset{+}{Br}CH_2CH_3)_2$	3.2 (m)		4.93 (t), $J = 7$	5.28 (q), $J = 7$	2.16 (t), $J = 7$	
$CH_2(CH_2\overset{+}{I}CH_3)_2$	3.2 (m)		4.60 (t), $J = 8$			3.80 (s)
$CH_2(CH_2\overset{+}{I}CH_2CH_3)_3$	3.2 (m)		4.60 (t), $J = 8$	4.83 (q), $J = 7$	2.30 (t), $J = 7$	

[a]Proton chemical shifts are referred to external capillary TMS. s, Singlet; d, doublet; t, triplet; q, quartet; qu, quintet; m, multiplet. J values are in hertz.
[b]Overlapping multiplet.

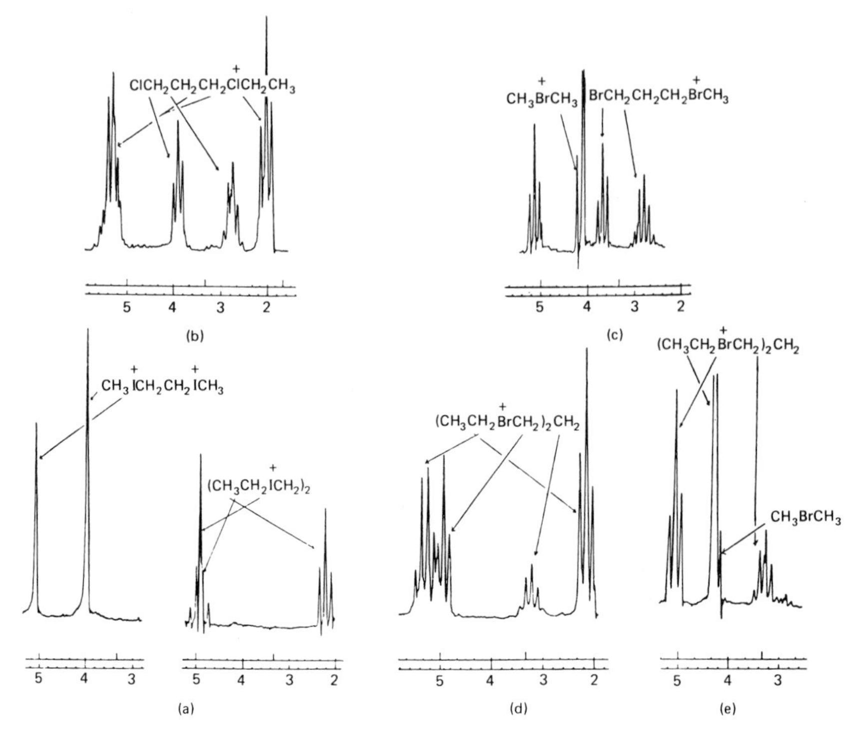

Figure 4. Pmr spectra of dialkyl alkylenedihalonium ions.

TABLE 14
Carbon-13 Nmr Data of Mono- and Dialkylated Diehaloalkanes[a]

Ion	$\delta_{^{13}C}$			
	$CH_3\overset{+}{X}-$	$-CH_2\overset{+}{X}-$	$-CH_2C\overset{+}{X}-$	$-CH_2CC\overset{+}{X}-$
$CH_3\overset{+}{I}CH_2\overset{+}{I}CH_3$	19.9 (160)	0.2 (183)		
$CH_3\overset{+}{I}CH_2I$	17.2 (159)	-24.1 (180)		
$CH_3\overset{+}{I}(CH_2)_3\overset{+}{I}CH_3$	11.1 (159)	33.4 (158)	30.1 (136)	
$CH_3\overset{+}{I}(CH_2)_3I$	10.3 (158)	41.3 (158)	33.0 (131)	6.2 (153)[c]
$CH_3\overset{+}{Br}CH_2Br$	40.4 (162)	50.7 (190)		
$CH_3\overset{+}{Br}(CH_2)_3\overset{+}{Br}CH_3$	39.9 (162)	60.2 (164)	28.7 (136)	
$CH_3\overset{+}{Br}(CH_2)_3Br$	38.4 (163)	67.4 (163)	32.0 (133)	31.6 (155)[d]

[a]Spectra were recorded on a Varian XL100FT spectrometer at −65°. The chemical shifts are refered to external TMS in SO_2 at −55°. J_{CH} values (in hertz) are given in parentheses.
[b]Assignments may be reversed.
[c]For CH_2I.
[d]For CH_2Br.

CHAPTER 4

Alkylarylhalonium Ions

4.1 PREPARATION AND NMR STUDY

Dence and Roberts[51] attempted to prepare the cyclopropylphenyliodonium ion from phenyliodoso chloride and cyclopropyllithium. However, they were unable to isolate the corresponding iodonium ion or any cyclopropylbenzene from

$$C_6H_5ICl_2 + \text{R-Li}$$
$$\rightarrow [C_6H_5\text{-}\overset{\oplus}{I}\text{-R}]Cl^{\ominus} \rightarrow C_6H_5I + RI$$
$$+ C_6H_5Cl + RCl$$
$$R = \text{cyclopropyl}$$

the reaction products. Thus the iodonium ion was not formed, even as an unstable reaction intermediate.

Perfluoroalkyliodoso trifluoroacetate reacts with aromatic compounds to give perfluoroalkylaryliodonium ions[52]

$$RI(OCOCF_3)_2 + C_6H_5R' \rightarrow R\text{-}\overset{\oplus}{I}\text{-}C_6H_4\text{-}R' \quad CF_3COO^{\ominus}$$

$$R = n\text{-}C_3F_7,\ n\text{-}C_6F_{13},\ C_6F_5$$

Anion-exchange reactions of these iodonium ions were carried out easily by using inorganic halides.

Alkylarylhalonium ions (other then perfluorininated derivatives) were first prepared by Olah and Melby.[53]

When a SO_2 solution of iodobenzene was added to a SO_2 solution of the $CH_3F\text{-}SbF_5$ complex (methyl fluoroantimonate) at $-78°$, a clear slightly colored solution resulted. The pmr spectrum of this solution at $-80°$ showed in addition to the excess methyl fluoroantimonate a methyl singlet at δ 3.80 and a multiplet aromatic region (7.7-8.3) with a peak area ratio of 3:5. The aromatic signals showed the same coupling pattern as that of iodobenzene in SO_2 but were deshielded by approximately 0.5 ppm. The fluorine-19 nmr spectrum of the solution showed only a very broad absorption centered at ϕ 112 (from CF_3CCl_3), which is characteristic of the SbF_6^- counterion. The species that

TABLE 15

Pmr Parameters of Alkylarylhalonium Ions[a]

Ion	CH_3X^+-	CH_3	$-CH_2-$	CH_3CX^+	Aromatic	Fluorine-19[b]
$C_6H_5Br^+CH_3$	4.45				7.5-7.9	
$C_6H_5I^+CH_3$	3.80				7.7-8.3	
$4\text{-}FC_6H_4I^+CH_3$	3.80				7.0-8.1	103.4
$4\text{-}FC_6H_4Br^+CH_3$	4.40				7.3-8.1	103.7
$4\text{-}CH_3C_6H_4I^+CH_3$	3.75	2.40			7.3-8.0	
$4\text{-}CH_3C_6H_4-Br^+CH_3$	4.35	2.40			7.4-7.9	
$C_6H_5I^+C_2H_5$			4.70	1.95	7.6-8.1	
$4\text{-}FC_6H_4I^+C_2H_5$			4.75	1.92	7.2-8.2	104.2
$C_6H_5Br^+C_2H_5$			5.30	1.90	7.7-8.2	
$4\text{-}FC_6H_4Br^+C_2H_5$			5.35	1.90	7.3-8.1	104.1

[a]Proton chemical shifts are from TMS in an external capillary tube. Spectra were recorded at −70° in SO_2 solution at 60 MHz. Relative peak areas were in agreement with expectation.

[b]Fluorine chemical shifts are from CF_3CCl_3 in an external capillary tube. The chemical shifts of 4-fluorobromobenzene and 4-fluoroiodobenzene in SO_2 solution are 114.8 and 113.9 ppm, respectively.

TABLE 16

Carbon-13 Chemical Shifts of Alkylarylhalonium Ions[a]

R_4–C_6H_2(ring positions 1–6; R_3 at 3, R_2 at 2)–$\overset{+}{X}$–CH_3 SbF_6^-

Ion	R_2	R_3	R_4	X	δ C-1	δ C-2	δ C-3	δ C-4	δ C-5	δ C-6	δ $\overset{+}{X}CH_3$	δ CH_3
$C_6H_5Br^+CH_3$	H	H	H	Br	127.6	134.6	132.7	135.5	132.7	134.6	50.0	
o-$CH_3C_6H_4Br^+CH_3$	CH_3	H	H	Br	128.9	142.2	135.0[b]	137.4[b]	133.8[b]	136.8[b]	49.2	23.4
m-$CH_3C_6H_4Br^+CH_3$	H	CH_3	H	Br	129.0	135.6	147.7	137.8	134.1	130.8	50.5	23.4
p-$CH_3C_6H_4Br^+CH_3$	H	H	CH_3	Br	125.4	136.6	133.7	148.8	133.7	136.6	50.7	23.4
o-$FC_6H_4Br^+CH_3$	F	H	H	Br	112.0 (22)	164.9 (25.4)	120.6 (19)	158.7 (7)	134.4	129.6	50.8	
m-$FC_6H_4Br^+CH_3$	H	F	H	Br	125.8 (10)	120.4 (26)	164.1 (25.4)	123.4 (19)	135.8 (10)	129.0	50.9	
p-$FC_6H_4Br^+CH_3$	H	H	F	Br	124.0	135.6 (9)	122.0 (22)	166.8	122.0 (22)	135.6 (9)	51.1	
$C_6H_5I^+CH_3$	H	H	H	I	106.4	137.8	133.9	133.9	133.9	137.8	20.5	
o-$CH_3C_6H_4I^+CH_3$	CH_3	H	H	I	112.7	144.4	134.7[b]	136.1[c]	131.8	140.0[c]	19.3	27.8
m-$CH_3C_6H_4I^+CH_3$	H	CH_3	H	I	106.7	138.5	145.4	135.3	133.9	136.1	20.3	23.0
p-$CH_3C_6H_4I^+CH_3$	H	H	CH_3	I	102.8	138.2	135.1	146.8	135.1	138.2	20.3	23.0
o-$C_6H_4I^+CH_3$	F	H	H	I	92.1 (22)	166.6 (25.0)	119.2 (22)	138.7 (10)	129.6	139.4	20.9	
m-$C_6H_4I^+CH_3$	H	F	H	I	104.6 (10)	125.5 (26)	163.8 (25.0)	123.1 (19)	135.8 (10)	154.6	21.1	
p-$C_6H_4I^+CH_3$	H	H	F	I	99.8	141.2 (10)	121.9 (22)	166.7 (25.7)	121.9 (22)	141.2 (10)	21.1	
$CH_3I^+C_6H_4I^+CH_3$	H	H	$^+ICH_3$	I	110.7	141.5	141.5	110.7	141.5	141.5	20.9	

[a] In parts per million from TMS. Hexafluoroantimonate salts in SO_2 at −60°, unless otherwise indicated. $J_{^{13}C\text{ and }F}$ values (in hertz) are given in parenthesis.

[b] Assignment tentative and would be reversed with respect to the ring positions.

accounts for the nmr data is the methylphenyliodonium ion ($CH_3I^+C_6H_5$). When bromobenzene and other aryl bromides or iodides were added in the same manner to methyl fluoroantimonate in SO_2, analogous spectra were obtained, indicating the formation of the corresponding methylarylbromonium ions.

$$C_6H_5X + CH_3F{\rightarrow}SbF_5 \ (C_2H_5F\text{-}SbF_5) \xrightarrow{SO_2} C_6H_5\overset{+}{X}{-}CH_3 \ (C_2H_5) \ SbF_6^-$$

$$X = Br, I$$

Likewise, the reaction of aryl bromides and iodides with ethyl fluoroantimonate in SO_2 gave the corresponding ethylarylhalonium ions. The structure of all alkylarylhalonium ions prepared was studied by nmr (hydrogen-1, carbon-13, and fluorine-19) spectroscopy. The proton and fluorine nmr data are summarized in Table 15. The proton chemical shifts of the methyl and ethyl protons are in good agreement with those of dialkylhalonium ions (Table 4). The aromatic protons are generally deshielded by approximately 0.5 ppm, corresponding to their halonium precursors. The fluorine chemical shifts of fluorinated alkylarylhalonium ions are slightly deshielded by ≈10 ppm from those of the corresponding precursors.

In order to study the trend of distribution of the positive charge and its delocalization into the aromatic ring, the Fourier-transform carbon-13 nmr shifts of a series of alkylarylhalonium ions (Table 16) were also determined.[26d]

In monosubstituted benzenes the carbon shieldings of most interest are the C-1 (ipso) carbon shielding because it occurs over the widest range of shifts,

TABLE 17
Aryl Carbon Shieldings in Phenylmethylhalonium Ions ($C_6H_5\overset{+}{X}CH_3$) and in Monosubstituted Benzenes (C_6H_5X)

		δ_C (ppm)[a]			
X		C-1	Ortho	Meta	Para
Br		123.4	132.1	131.0	127.8
$^+BrCH_3$		127.6	134.6	132.7	135.5
	Δ	4.2	2.5	1.7	7.7
I		96.8	139.0	131.7	129.8
$^+ICH_3$		106.4	137.8	133.9	133.9
	Δ	9.6	-1.2	2.2	4.1

[a]Referred to TMS.

and the C-4 (para) carbon shielding because, apart from a few exceptions, it reflects the electron-withdrawing or electron-donating ability of the substituent. These shifts in bromo- and iodobenzene for the methylphenylbromonium and iodonium ions are shown in Table 17).

The data show that upon methylation of the ring halogen, the largest downfield shifts occur for the ipso and para carbons. This suggests the following resonance structures for alkylarylhalonium ions:

$$\text{C}_6\text{H}_5\text{-X}^+\text{-R}_1 \longleftrightarrow {}^+\text{C}_6\text{H}_5\text{=X-R}_1 \longleftrightarrow {}^+\text{C}_6\text{H}_5\text{=X-R} \quad \text{etc.}$$

The charge delocalization into the ortho and para positions is reflected by the deshielding effect on these carbons (Table 17).

These changes are too large to be solvent effects alone, and suggest that the CH_3X^+ group is somewhat electron-withdrawing compared with X. The smaller change in the para-carbon shift for iodine reflects the lesser ability of its lone-pair electrons, compared with those of bromine, to interact with the π electrons of the benzene ring. The para-carbon shift change in both cases is smaller than that observed (10-15 ppm) on protonation of monosubstituted benzenes (e.g., benzoic acid, benzaldehyde, nitrobenzene) and shows the lesser ability of the halogen lone pairs to interact with the π electrons of the benzene ring.

Comparison of the relevant values in Table 16 and those of haloalkanes and -arenes shows that the α-substitutent effects in alkyl and aryl halides are significantly different. This is true for many substituents other than halogens and most likely is attributable to the greater polarizability of the phenyl ring system. A similar difference, only larger, is observed for the α-carbon shifts of aryl- and alkylmethylhalonium ions ($\delta_{C_{ipso}}$ and δ_{CH_3}). The larger difference results from a much larger deshielding of the methyl resonance on methylation of methyl iodide (29.9 ppm for iodine and 27.3 ppm for bromine) than of the ipso-carbon resonance on methylation of aryl halides (6 ppm for iodine and 4.2 ppm for bromine).

The β-substituent effect is approximately the same for alkyl and aryl iodides (10-11 ppm) and is unaffected by methylation of the iodine atom. The situation is slightly different in bromobenzene, where this effect is 5.8 ppm, or about one-half that in alkyl bromides. Placing a positive charge on the bromine by methylation results slight in deshielding from aryl bromides and slight shielding from alkyl bromides.

The effect of individual substituents in many disubstituted benzenes appears to be additive, provided that substituents are meta or para to each other. This is shown by the ipso-carbon shielding of p-CH_3I^+–, p-CH_3–, and p-fluoromethylphenyliodonium ions, which increase to higher field in that order, in agreement with the observed para-substituent effects of CH_3I^+, methyl, and fluorine (+5.1, −3.4, and −4.4 ppm, respectively). The same trend is found in para-substituted methylphenylbromonium ions.

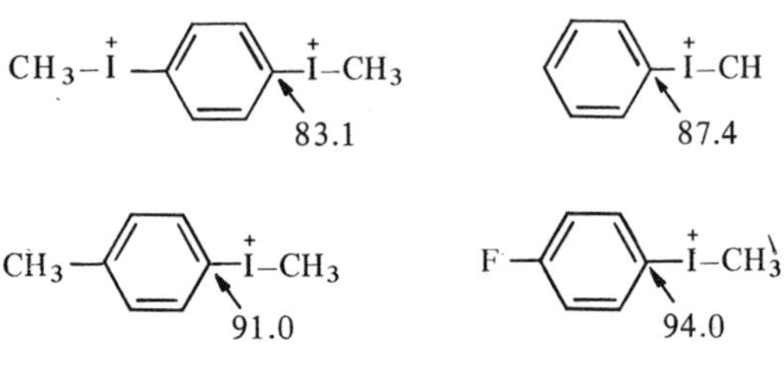

$CH_3-\overset{+}{I}-C_6H_4-\overset{+}{I}-CH_3$ 83.1

$C_6H_5-\overset{+}{I}-CH_3$ 87.4

$CH_3-C_6H_4-\overset{+}{I}-CH_3$ 91.0

$F-C_6H_4-\overset{+}{I}-CH_3$ 94.0

Consistent with the much smaller meta-substituent effects found in monosubstituted benzenes, the ipso-carbon shielding in meta-substituted methylphenylhalonium ions is almost constant. Trends similar to those mentioned above have been observed in m- and p-CH_3, fluorine-, and C=O-substituted benzoyl cations.

It has been noted that methyl carbon shielding in substituted toluenes is essentially independent of the nature of the meta and para substituents, but that ortho substituents produce upfield shifts, except in the case of iodine. Data for CH_3I^+ and CH_3Br^+ substituents are consistent with these observations, since in the meta and para derivatives of toluene, δ_{CH_3} is 170.6 ± 0.3 ppm [δ_{CH_3} (toluene) = 172.6] and δCH_3 in the o-CH_3I^+ derivative of toluene is 166.0. The methyl resonance in the corresponding ortho-bromo compound, unlike iodine, is essentially unchanged from that in the meta and para derivatives.

The shielding of a methyl group bonded directly to halogen in arylalkylhalonium ions is unaffected by meta or para ring substituents but is affected slightly by an ortho ring substituent. It is of interest that these methyl group shifts in methylarylhalonium ions are deshielded by 11-13 ppm from the corresponding signals in alkylmethylhalonium ions, implying that ArX^+ is more electronegative than CH_3X^+, since this difference is too large to be accounted for by anisotropy contributions from the phenyl ring. Perhaps this may be regarded as additional evidence for charge delocalization into the phenyl ring, except that the deshielding is of the same magnitude for both iodine and bromine.

The reaction of chlorobenzene and fluorobenzene with methyl fluoroantimonate in SO_2 does not result in the formation of alkylarylhalonium ions, but instead ring-methylated and sulfonylmethylated products are found (the latter preferably in the para position). Methyl fluoroantimonate apparently also methylates SO_2, forming an effective sulfonylating agent. When chlorobenzene

$$CH_3F\text{-}SbF_5 + SO_2 \rightarrow [CH_3\text{-}\overset{+}{SO_2}]SbF_6^- \xrightarrow{C_6H_5Cl(F)} [\text{Cl(F)-}C_6H_5^+(H)(SO_2CH_3)] \xrightarrow{-H^+} \text{Cl(F)-}C_6H_4\text{-}SO_2CH_3$$

is added to methyl fluoroantimonate in SO_2ClF, sulfonylmethylation does not occur, while methylation occurs on the aromatic ring to give chlorotoluenes and chloroxylenes.

4.2 RELATIVE STABILITY, ALKYLATING ABILITY, AND ROLE IN FRIEDEL-CRAFTS REACTIONS

All the studied methylarylhalonium ions are stable up to $-20°$. When a SO_2 solution of methylphenylbromonium ions is heated in a sealed tube to $0°$, transformation readily occurs to give a mixture of bromoxylenes. The SO_2 solution of methylphenyliodonium ions is considerable more stable. However, after 15 hr at room temperature in a sealed tube, transformation to C-methylated products occurs. Transformation of ethylphenylbromonium ions occurs readily at $-70°$ to give C-methylated products.[12]

Transformation of alkylarylhalonium ions (at higher temperatures) to the C-methylated products may play a role in Friedel-Crafts alkylation of halobenzenes. However, studies have shown that the transformation of alkylarylhalonium ions to C-alkylated haloaromatics is intermolecular rather than intramolecular.

Alkylarylhalonium ion were shown to be good general alkylating agents not only for π-donor aromatic systems, but also for *n* bases. They alkylate ethers, amines, and methyl halides to give the corresponding onium ions. The fact that alkylarylhalonium ions irreversibly react with methyl halides to give dimethylhalonium ions shows that methyl halides have a greater affinity for incipient methyl cations than halobenzenes.

4.3 DIALKYL PHENYLENEDIHALONIUM IONS

The alkylation of dihaloarenes with methyl and ethyl fluoroantimonate in SO_2 solution at low temperature was studied. Mono- or dihalonium ion formation takes place under various conditions, depending on the nature of the substrates as well as the alkylating reagents.[13] Data are summarized in Table 18. The unusually strong donor ability of iodine leads to formation of three diiodonium ions from the corresponding isomeric diiodobenzenes (Figure 5). However, only *p*-dibromobenzene (of the three isomeric dibromobenzenes) was dialkylated to give a dibromonium ion with methyl fluoroantimonate in

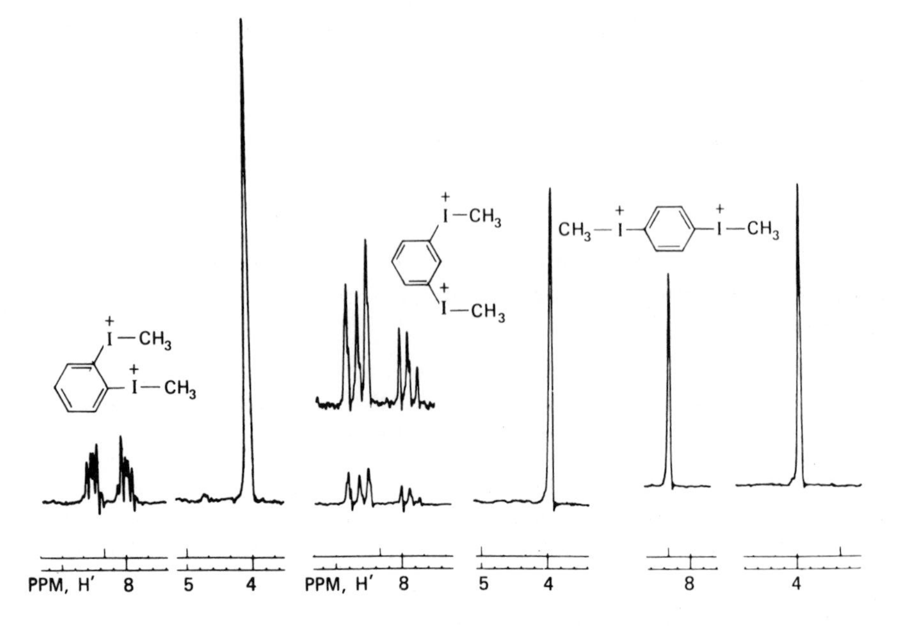

Figure 5. Pmr spectra of dimethylated *o*-diiodobenzene (left), *m*-diiodobenzene (middle), and *p*-diiodobenzene (right).

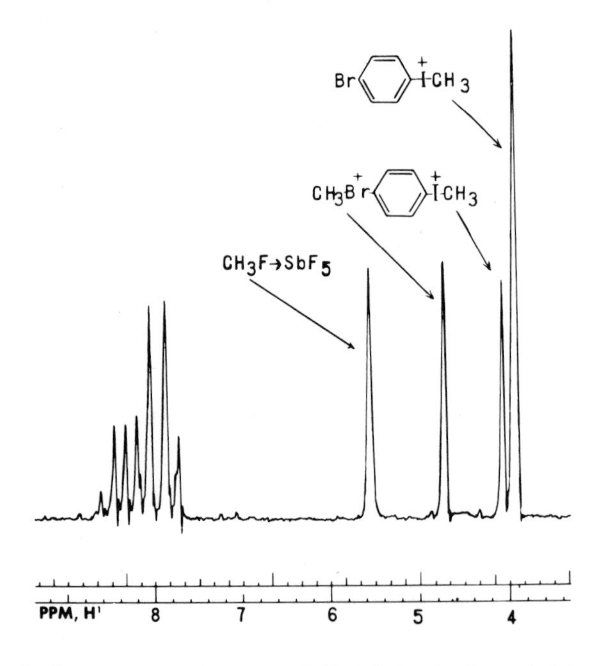

Figure 6. Pmr spectrum of mono- and dimethylated *p*-bromoiodobenzene.

TABLE 18

Alkylation of Di- and Trihaloarenes with Methyl and Ethyl Fluoroantimonate

Di- and tri-haloarene	Alkylating reagent, $RF\text{-}SbF_5\text{-}SO_2$	Halonium ion formed	Remarks
$I-C_6H_4-I$ (*o*-, *m*-, *p*-)	$CH_3F\text{-}SbF_5\text{-}SO_2$ (1 mole)	$I-C_6H_4-I^+CH_3$	
	$CH_3F\text{-}SbF_5\text{-}SO_2$ (excess)	$CH_3I^+-C_6H_4-I^+CH_3$	
	$C_2H_5F\text{-}SbF_5\text{-}SO_2$ (1 mole)	$I-C_6H_4-I^+CH_2CH_3$	
	$C_2H_5F\text{-}SbF_5\text{-}SO_2$ (excess)	$CH_3CH_2I^+-C_6H_4-I^+CH_2CH_3$	
$Br-C_6H_4-Br$ (*o*-, *m*-)	$CH_3F\text{-}SbF_5\text{-}SO_2$ (excess)	$Br-C_6H_4-Br^+CH_3$	
$Br-C_6H_4-Br$ (*p*-)	$CH_3F\text{-}SbF_5\text{-}SO_2$ (excess)	$Br-C_6H_4-Br^+CH_3 \rightleftharpoons CH_3Br^+-C_6H_4-Br^+CH_3$	Only monohalonium ion observed at -10°
$Br-C_6H_4-Br$ (*o*-, *m*-, *p*-)	$C_2H_5F\text{-}SbF_5\text{-}SO_2$ (excess)	Mixture of ring ethylation products	

TABLE 18 *(continued)*

Di- and tri-haloarene	Alkylating reagent, $RF\text{-}SbF_5\text{-}SO_2$	Halonium ion formed	Remarks
$Br-C_6(CH_3)_4-Br$	$CH_3F\text{-}SbF_5\text{-}SO_2$ (excess)	$CH_3Br^+-C_6(CH_3)_4-Br^+CH_3$	
	$C_2H_5F\text{-}SbF_5\text{-}SO_2$ (excess)	$CH_3CH_2Br^+-C_6(CH_3)_4-Br^+CH_2CH_3$	
$I-C_6F_4-I$	$CH_3F\text{-}SbF_5\text{-}SO_2$ (excess)	$CH_3I^+-C_6F_4-I^+CH_3$ + $I-C_6F_4-I^+CH_3$	
	$C_2H_5F\text{-}SbF_5\text{-}SO_2$ (excess)		Halonium ions formed are not soluble enough to be identified
$Br-C_6F_4-Br$	$CH_3F\text{-}SbF_5\text{-}SO_2$ (excess) or $C_2H_5F\text{-}SbF_5\text{-}SO_2$ (excess)	No reaction	
$Br-C_6H_4-I$	$CH_3F\text{-}SbF_5\text{-}SO_2$ (excess)	$CH_3Br^+-C_6H_4-I^+CH_3$ + $Br-C_6H_4-I^+CH_3$	

I, I, I	$C_2H_5F\text{-}SbF_5\text{-}SO_2$ (excess)	Mixture of ring ethylation products	
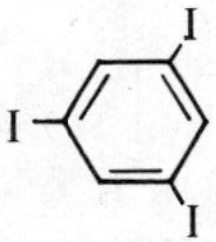	$CH_3F\text{-}SbF_5\text{-}SO_2$ (excess)	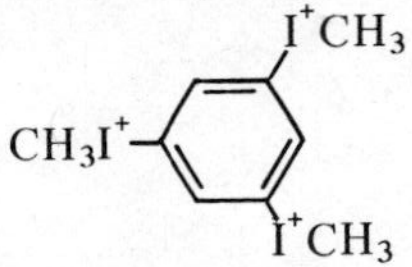 I^+CH_3, CH_3I^+, I^+CH_3	
	$C_2H_5F\text{-}SbF_5\text{-}SO_2$ (excess)	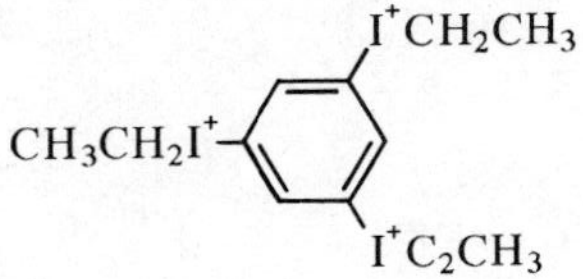 $I^+CH_2CH_3$, $CH_3CH_2I^+$, $I^+C_2CH_3$	
	$CH_3F\text{-}SbF_5\text{-}SO_2$ (1 mole) or $C_2H_5F\text{-}SbF_5\text{-}SO_2$ (1 mole)	No reaction	1,3,5-Triiodobenzene does not dissolve in the alkylating reagents
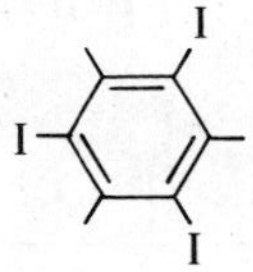 I, I, I	$CH_3F\text{-}SbF_5\text{-}SO_2ClF\text{-}SO_2$ (excess)	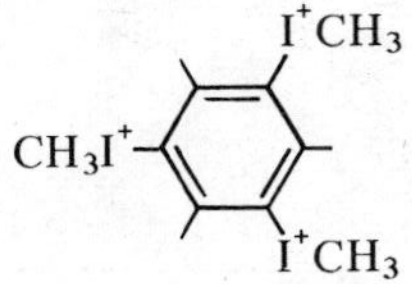 I^+CH_3, CH_3I^+, I^+CH_3	
	$CH_3CH_2F\text{-}SbF_5\text{-}SO_2$ (excess)	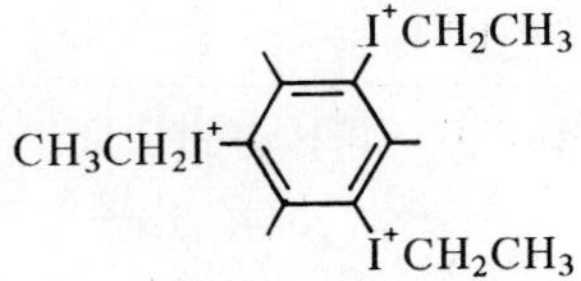 $I^+CH_2CH_3$, $CH_3CH_2I^+$, $I^+CH_2CH_3$	
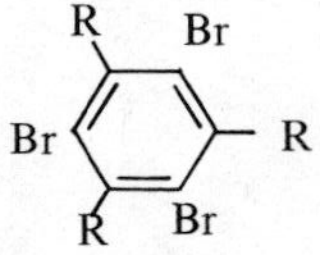 R, Br, Br, R, R, Br R = H, F, CH_3	$CH_3F\text{-}SbF_5\text{-}SO_2$ (excess)	No reaction	1,3,5-Tribromobenzene does not dissolve in the alkylating reagent

TABLE 19

Pmr Parameters of Mono-, Di-, and Trialkylated Halobenzenes[a]

Ion	δ XCH_3	δ CH_3	δ CH_3CX	δ CH_2X	δ Aromatic
$o\text{-}(CH_3\overset{+}{I})_2C_6H_4$	4.10 (s)				8.0-8.8 (m)
$o\text{-}(C_2H_5\overset{+}{I})C_6H_4$			2.23 (t), $J = 7$	5.23 (q), $J = 7$	8.1-8.7 (m)
$m\text{-}(CH_3\overset{+}{I})_2C_6H_4$	3.97 (s)				7.7-8.9 (m)
$m\text{-}(C_2H_5\overset{+}{I})_2C_6H_4$			2.15 (t), $J = 7.5$	5.06 (q), $J = 7.5$	7.8-8.7 (m)
$p\text{-}(CH_3\overset{+}{I})_2C_6H_4$	3.95 (s)				8.35 (s)
$p\text{-}(C_2H_5\overset{+}{I})_2C_6H_4$			2.14 (t), $J = 7.5$	5.02 (q), $J = 7.5$	8.35 (s)
$o\text{-}BrC_6H_4\overset{+}{Br}CH_3$	4.55 (s)				7.6-8.3 (m)
$m\text{-}BrC_6H_4\overset{+}{Br}CH_3$	4.48 (s)				7.5-8.2 (m)
$p\text{-}BrC_6H_4\overset{+}{Br}CH_3$	4.53 (s)				7.95 (s)
$p\text{-}(CH_3\overset{+}{Br})_2C_6H_4$	4.56 (s)				8.38 (s)
$p\text{-}CH_3\overset{+}{Br})_2C_6(CH_3)_4$	4.40 (s)	2.85 (s)			
$p\text{-}(C_2H_5\overset{+}{Br})_2C_6(CH_3)_4$		2.80 (s)	2.05 (t), $J = 7$	5.30 (q), $J = 7$	

$p\text{-}IC_6H_4\overset{+}{I}CH_3$	4.20 (s)				
$p\text{-}(CH_3\overset{+}{I})_2C_6F_4$	4.31 (s)				7.7-8.2 (m)
$p\text{-}BrC_6H_4\overset{+}{I}CH_3$	3.90 (s)				7.7-8.2 (m)
$p\text{-}CH_3\overset{+}{Br}C_6H_4\overset{+}{I}CH_3$	4.05 (s)				8.2-8.7 (m)
	4.65 (s)				
$1,3,5(CH_3\overset{+}{I})_3C_6(CH_3)_3$	4.00 (s)	3.65 (s)			
$1,3,5(C_2H_5\overset{+}{I})_3C_6(C_2H_5)_3$		3.60 (s)	2.18 (t), $J = 7$	5.20 (q), $J = 7$	
$1,3,5(CH_3\overset{+}{I})_3C_6H_3$	4.00 (s)				8.38 (s)
$1,3,5(C_2H_5\overset{+}{I})_3C_6H_3$			2.12 (t), $J = 7$	5.02 (t), $J = 7$	8.30 (s)

[a] From TMS in external capillary tube. Spectra were recorded at -70° in SO_2 solution at 60 MHz. s, Singlet; t, triplet; q, quartet; m, multiplet.

SO_2 solution at $-78°$. *p*-Bromoiodobenzene was both mono- and dimethylated under the same conditions (Figure 6). The structural proof of mono- and

$RF\text{-}SbF_5\text{-}SO_2, -78°$

X = Br, I
R = CH_3, C_2H_5

dialkylated halobenzenes as real dialkyl phenylenedihalonium ions was based on pmr studies (Table 19). Dichlorobenzenes, however, showed only ring alkylation with methyl and ethyl fluoroantimonates without giving observable mono- or dihalonium ions. Dialkyl phenylenedihalonium ions prepared at higher temperature decompose, again apparently by intermolecular processes, to give ring-alkylated haloarenes.

The substituents on the aromatic ring also affect the ease of formation of dihalonium ions. For example, 2,3,5,6-tetramethyldibromobenzene was dialkylated with methyl and ethyl fluoroantimonate in SO_2 solution at $-78°$, while 2,3,5,6-tetrafluorodibromobenzene was inert to the same alkylating reagents. In contrast, 2,3,5,6-tetrafluorodiiodobenzene was found to be dimethylated with excess methyl fluoroantimonate in SO_2 solution at $-78°$.

$RF\text{-}SbF_5\text{-}SO_2, -78°$

R = CH_3, C_2H_5
X = Br, I
R′ = CH_3, F

R = R′ = CH_3; X = Br
R = R′ = C_2H_5; X = Br
R = CH_3; R′ = F; X = I

4.4 TRIALKYL PHENYLENETRIIODONIUM IONS

1,3,5-Triiodobenzene and 2,4,6-triiodomesitylene were trialkylated with excess methyl and ethyl fluoroantimonates in SO_2 (SO_2ClF) solution to the corresponding trihalonium ions.[13] However, alkylation of 1,3,5-tribromobenzene, 2,4,6-

$RF\text{-}SbF_5\text{-}SO_2, -78°$

$R = CH_3;\ R' = H$
$R = C_2H_5;\ R' = H$
$R = R' = CH_3$
$R = C_2H_5;\ R' = CH_3$

tribromomesitylene, and 1,3,5-tribromotrifluorobenzene under similar conditions failed, since the starting materials were insoluble in the alkylating reagents. A study of the alkylation of trihaloarenes with methyl and ethyl fluoroantimonates is summarized in Table 17. The structure of the trihalonium ions is in agreement with pmr and cmr data (Tables 18 and 19).

The unusual donor ability of iodine toward electrophiles is reflected in the formation of dihalonium ions, even in the case of diiodomethane and *o*-diiodobenzene. The formation of trimethyl phenylenetriiodonium ions further reveals the ability of iodine to accommodate positive charge. Dialkyl alkylenedibromonium ions were formed only when dibromopropanes were treated with methyl (or ethyl)fluoroantimonate solution at low temperature. Methylation of dibromobenzenes results in the formation of either monomethylated species or a mixture of monomethylated and dimethylated species. It was not found possible to trimethylate tribromobenzenes. Dialkyl phenylenedichloronium ions have so far not been observed. These results reveal that the ease of halonium ion or dihalonium ion formation in alkylarylhalonium ions also decreases in the order $\diagdown\overset{+}{I}\diagup < \diagdown\overset{+}{Br}\diagup < \diagdown\overset{+}{Cl}\diagup$.

CHAPTER 5

Diarylhalonium Ions

In contrast to dialkylhalonium ions and alkylarylhalonium ions, diarylhalonium ions are considerably more stable. This is particularly the case for diaryliodonium ions, which have been known for 80 years.[1] It is interesting, however, to compare the discovery and assumed significance of these ions with that of the related carbocations, that is, triarylcarbenium ions (triarylmethyl cations). The latter were also discovered at about the turn of the century, but were considered only a specific class of organic cations limited exclusively to the highly stabilized triarylmethyl systems. The general significance of carbocations as intermediates in electrophilic reactions was not recognized until many years later, when it became evident that they are intermediates in all electrophilic organic reactions. Subsequently, methods were developed to prepare practically any conceivable type of carbocations under stable ion conditions. Halonium ions represent a somewhat similar case. Diaryliodonium (or to a lesser extent diarylbromonium and -chloronium ions) were at first considered a specific class of highly stabilized halonium ions. No relationship or significance was attached to these ions until many years later, when it was realized that many other types of halonium ions (such as dialkyl-, alkylaryl-, and alkylenehalonium ions) can exist, and the general significance of halonium ions in electrophilic halogenation reactions of organic compounds was pointed out.

5.1 DIARYLIODONIUM IONS

5.1.1 Preparation and Isolation

A variety of synthetic methods for diaryliodonium ions has been developed since Hartman and Meyer discovered them by preparing *p*-iododiphenyliodonium bisulfate salts from iodosobenzene in concentrated H_2SO_4 in 1894[1].

$$2C_6H_5IO + H_2SO_4 \rightarrow I\text{-}C_6H_4\text{-}\overset{\oplus}{I}\text{-}C_6H_5 \; HSO_4^{\ominus} + H_2O + [O]$$

The synthetic methods for preparation of diaryliodonium ions can be divided into those for symmetric and unsymmetric diaryliodonium ions.[11]

Symmetric Diaryliodonium Ions

In concentrated H_2SO_4 diaryliodonium ions are obtained from iodyl sulfate, which is synthesized by decomposing iodic acid by heating in concentrated H_2SO_4, and aromatic compounds in good yield.[54]

$$4\text{Ar-H} + (\text{IO})_2\text{SO}_4 + \text{H}_2\text{SO}_4 \rightarrow 2\text{Ar}_2\text{I}^{\oplus}\text{HSO}_4^{\ominus} + 2\text{H}_2\text{O}$$

By using iodine(III) trifluoroacetate, prepared from iodine,[55] trifluoroacetic acid, and fuming nitric acid, diaryliodonium ions are obtained from aromatic compounds and trifluoroacetic acid in acetic anhydride. The yields are, however, generally not too high.

$$2\text{Ar–H} + \text{I(OOCCF}_3)_3 + \text{HX} \rightarrow \text{Ar}_2\text{I}^{\oplus}\text{X}^{\ominus} + 3\text{CF}_3\text{CO}_2\text{H}$$

Aromatics react with alkali iodates in acetic acid-acetic anhydride-H_2SO_4 solution to give diaryliodonium ions.[54] This method is very useful for the isolation of soluble salts such as diaryliodonium chlorides, nitrates, and bisulfates.

$$2\text{Ar–H} + \text{KIO}_3 + \text{H}_2\text{SO}_4 + 2(\text{CH}_3\text{CO})_2\text{O}$$
$$\rightarrow \text{Ar}_2\text{I}^{\oplus}\text{HSO}_4^{\ominus} + 4\text{CH}_3\text{CO}_2\text{H} + \text{KHSO}_4 + [\text{O}]$$

Unsymmetric Diaryliodonium Ions

The most convenient preparation involves the condensation of iodosoarenes or iodoxyarenes with aromatics in the presence of H_2SO_4. In the case of an

$$\text{ArIO} + \text{Ar}'\text{H} + \text{H}_2\text{SO}_4$$
$$\rightarrow \text{Ar–I}^{\oplus}\text{-Ar}'\ \text{HSO}_4^{\ominus} + \text{H}_2\text{O}$$ [54,56]

$$\text{ArI(OAc)}_2 + \text{Ar}'\text{H} + \text{H}_2\text{SO}_4$$
$$\rightarrow \text{ArI}^{\oplus}\text{–Ar}'\ \text{HSO}_4^{\ominus} + 2\ \text{AcOH}$$ [54,57]

$$\text{ArIO}_2 + \text{Ar}'\text{H} + \text{H}_2\text{SO}_4$$
$$\rightarrow \text{ArI}^{\oplus}\text{Ar}'\ \text{HSO}_4^{\ominus} + \text{H}_2\text{O} + [\text{O}]$$ [54]

iodosoarene bearing an electron-withdrawing group, such as a nitro group, the reaction is carried out in H_2SO_4 to give diaryliodonium ions in good yield. In the case of electron-releasing groups, the reaction proceeds in acetic anhydride, giving good yield.

Iodoarenes can be reacted with arenes by *in situ* oxidation of the iodoarene to iodosoarenes which, without isolation, can be coupled to form diaryliodium ions.[54,58] Potassium persulfate and barium oxide are most useful as oxidants

$$\mathrm{ArI + Ar'H + K_2S_2O_8 + H_2SO_4 \rightarrow ArI^{\oplus}Ar'\ HSO_4^{\ominus} + 2KHSO_4}$$

$$\mathrm{ArI + Ar'H + BaO_2 + 2H_2SO_4 \rightarrow ArI^{\oplus}{-}AR'\ HSO_4^{\ominus} + BaSO_4 + 2H_2O}$$

of iodoarenes. This method is suitable for iodoarenes bearing electron-withdrawing groups.

$$\mathrm{ArI + K_2S_2O_8 + 2H_2SO_4 \rightarrow ArI(OSO_3H)_2 + 2KHSO_4}$$

$$\mathrm{ArI(OSO_3H)_2 + Ar'H \rightarrow ArI^{\oplus}Ar'\ HSO_4^{\ominus} + H_2SO_4}$$

$$\mathrm{ArI + BaO_2 + 3H_2SO_4 \rightarrow ArI(OSO_3H)_2 + 2H_2O + BaSO_4}$$

$$\mathrm{ArI(OSO_3H)_2 + Ar'H \rightarrow ArI^{\oplus}Ar'\ HSO_4^{\ominus} + H_2SO_4}$$

Aryllithium compounds react with aryliodoso chloride at low temperatures to give diaryliodonium ions.[59] However, the yields are not good. Usually,

$$\mathrm{ArICl_2 + ArLi \rightarrow ArI^{\oplus}Ar^{+}\ Cl^{\oplus} + LiCl}$$

trivalent iodine compounds such as phenyliodoso chloride and iodosobenzene are used in the synthesis of diaryliodonium ions. However, Nesmeyanov et al.[60] have also reported the synthesis of diaryliodonium ions from aryldiazonium salts and iodoarenes. In this reaction aryl cations can be formed by decomposition of the aryldiazonium salt, which react with the iodoarene to give the corresponding diaryliodonium ions. Except for cyclic iodonium ions, however,

$$\mathrm{ArN_2^{\oplus}BF_4^{\ominus} + Ar'I \xrightarrow{\Delta} Ar{-}I^{\oplus}{-}Ar'\ BF_4^{\oplus}}$$

Ar = $\mathrm{EtO_2C{-}C_6H_4{-}}$ (para), 3-nitrophenyl ($\mathrm{NO_2}$)

$$Ar' = C_6H_5-$$

diaryliodonium ions have not been prepared from iodoarenes and aryl cations, formed via the decomposition of diazonium salts, in satisfactory yields. Interestingly, Miller and Hoffman[61] have reported the electrochemical synthesis

$$2C_6H_5I \longrightarrow C_6H_5-\overset{\oplus}{I}-C_6H_4-I \quad \overset{\ominus}{ClO_4}$$

$$C_6H_5I + C_6H_5OCH_3 \longrightarrow C_6H_5-\overset{\oplus}{I}-C_6H_4-OCH_3 \quad ClO_4^{\ominus}$$

of diaryliodonium ions. The cation radical of iodobenzene is first formed and reacts with the aromatic compound to give the corresponding diaryliodonium

$$ArI \xrightarrow{-e} ArI\cdot^{\oplus}$$

$$ArI^{\oplus} + C_6H_5R \rightarrow ArI^{\oplus}(H)C_6H_5R \xrightarrow{-e} ArI^{\oplus}-C_6H_4-R + H^{\oplus}$$

ion. Beringer et al.[62] have reported the synthesis of diaryliodonium ions from aryllithiums and *trans*-chlorovinyliodoso dichloride. Ions prepared are summarized in Table 20. Concerning the mechanism of the reaction they suggested

TABLE 20
Preparation of Diaryliodonium Ions from Aryllithiums and *trans*-Chlorovinyliodoso Dichloride

Diaryliodonium ion	Yield (%)
$Ph_2I^{\oplus}I^{\ominus}$	100
$(p\text{-Tolyl})_2I^{\oplus}I^{\ominus}$	73
$(2\text{-Naphtlyl})_2I^{\oplus}Cl^{\ominus}$	55
$(9\text{-Anthranyl})_2I^{\oplus}Cl^{\ominus}$	32
Dibenziodophenium $Cl^{\ominus}$	39

$$2\text{Ar–Li} + \text{H(Cl)C=C(H)ICl}_2 \rightarrow \text{Ar–}\overset{\oplus}{\text{I}}\text{–Ar}\,\text{Cl}^{\ominus} + \text{CH}\equiv\text{CH} + 2\text{LiCl}$$

the formation of diaryl-*trans*-β-chloroethenyliodine as the intermediate, which decomposes to give the corresponding diaryliodonium ion and acetylene.

$$\text{H(Cl)C=C(H)ICl}_2 + \text{Ar–Li} \xrightarrow{-\text{LiCl}} \text{H(Cl)C=C(H)}\overset{\oplus}{\text{I}}\text{–Ar}\ \text{Cl}^{\ominus} \xrightarrow{\text{Ar–Li}} \text{H(Cl)C=C(H)I(Ar)}_2 \longrightarrow \text{Ar}_2\text{I}^{\oplus}\text{Cl}^{\ominus} + \text{CH}\equiv\text{CH}$$

Phenylpentafluorophenyliodonium ions were prepared from phenyliodoso dichloride and pentafluorophenyllithium in good yield.[63]

$$C_6H_5ICl_2 + C_6F_5Li \xrightarrow[-78^{\circ}C]{} C_6H_5\text{–}\overset{\oplus}{I}\text{–}C_6F_5\ Cl^{\ominus} \xrightarrow{CF_3CO_2H} C_6H_5\overset{\oplus}{I}\text{–}C_6F_5\ CF_3CO_2^{\ominus}$$

Heteroaryliodonium Ions

Beringer et al.[57] synthesized 2,2′-dithienyliodonium ions from thiophene and potassium iodate in acetic acid-acetic anhydride-concentrated H_2SO_4

$$2\ \text{(thiophene)} + KIO_3 + H_2SO_4 + 2Ac_2O \rightarrow \text{(2-thienyl)–}\overset{\oplus}{I}\text{–(2-thienyl)}\ HSO_4^{\ominus} + 4AcOH + [O]$$

solution. Subsequently, they also synthesized these iodonium ions from 2-thienyl lithium and *trans*-β-chloroiodosochloroethylene in good yield.[64]

$$2\ (2\text{-thienyl})\text{-Li} + \text{Cl(H)C=C(H)ICl}_2 \longrightarrow (2\text{-thienyl})_2\text{I}^{\oplus}\ \text{Cl}^{\ominus} + 2\text{LiCl} + \text{CH}\equiv\text{CH}$$

The 2,2′-difurfuryliodonium ion has also been synthesized from 2-furfuryl lithium and *trans*-β-chloroiodosochloroethylene by the same method.

$$2\ (2\text{-thienyl})\text{-Li} + \text{Cl(H)C=C(H)ICl}_2 \longrightarrow (2\text{-thienyl})_2\text{I}^{\oplus}\ \text{Cl}^{\ominus} + 2\text{LiCl} + \text{CH}\equiv\text{CH}$$

From 2-pyridyllithium and *trans*-β-chloroidosochloroethylene, the corresponding iodonium ion could not be obtained.

$$(2\text{-pyridyl})\text{-Li} + \text{Cl(H)C=C(H)ICl}_2 \not\longrightarrow (2\text{-pyridyl})_2\text{I}^{\oplus}\ \text{Cl}^{\ominus} + 2\text{LiCl} + \text{CH}\equiv\text{CH}$$

3-Iodosopyridine was allowed to react with an aryl compound to give 3-pyridylaryliodonium ions.[65]

$$(3\text{-pyridyl})\text{IO} + \text{ArH} \longrightarrow (3\text{-pyridyl})\overset{\oplus}{\text{I}}\text{-Ar}\ \text{X}^{\ominus}$$

Ar = CH_3O–C₆H₄–, C_6H_5O–C₆H₄–, 2-thienyl

Iodonium Ions Bearing Olefinic or Acetylenic Ligands

trans-Chlorovinylmercuric chloride reacts with iodine trichloride to give the double salt of bis(*trans*-2-chlorovinyl)iodonium chloride in 6.0% yield, which is treated wtih hydrogen sulfide to give a free iodonium ion.[66] By treating

$$\underset{Cl}{\overset{H}{}}C{=}C\underset{H}{\overset{HgCl}{}} \xrightarrow{ICl_3} (ClCH{=}CH)_2I^{\oplus}Cl^{\ominus}\ 2HgCl_2$$

$$\xrightarrow{ArICl_2} ClCH{=}CH{-}\overset{\oplus}{I}{-}Ar\quad HgCl_3^{\ominus}$$

trans-chlorovinylmercuric chloride with aryliodoso chloride, aryl(*trans*-chlorovinyl)iodonium ion was formed in 31% yield,[67] which eliminated acetylene on treatment with alkali solution. This iodonium ion was synthesized also in the reaction of *trans*-β-chloroiodosochloroethylene with phenyltin trichloride.

$$C_6H_5SnCl_3 + \underset{Cl}{\overset{H}{}}C{=}C\underset{H}{\overset{ICl_2}{}} \xrightarrow[X_1\ =\ 31\%]{} ClCH{=}CH{-}\overset{\oplus}{I}{-}C_6H_5\ Cl^{\ominus} + SnCl_4$$

Phenylethynyllithium reacts with phenyliodoso chloride to give phenylethynyl, phenyliodonium chloride which is treated with sodium fluoroborate to give the β,β-phenylchlorovinyl, phenyliodonium ion.[68] Nesmeyanov et al.[69] have report-

$$C_6H_5{-}C{\equiv}C{-}Li + C_6H_5ICl_2 \longrightarrow C_6H_5{-}C{\equiv}C{-}\overset{\oplus}{I}{-}C_6H_5\ Cl^{\ominus}$$

$$\xrightarrow{NaBF_4\text{-}H_2O} \underset{Cl}{\overset{C_6H_5}{}}C{=}C\underset{\overset{\oplus}{I}-C_6H_5}{\overset{H}{}}\quad BF_4^{\ominus}$$

ed the synthesis of the iodonium ion bearing a β-styryl group as a ligand from β-phenylvinyltin trichloride or β-phenylvinylmercuric bromide and phenyliodoso chloride, although in very low yield. The reactivity of the vinyl groups to

$$C_6H_5CH{=}CH{-}SnCl_3 + C_6H_5ICl_2 \rightarrow C_6H_5\overset{\oplus}{I}CH{=}CH{-}C_6H_5\quad I^{\ominus}$$

$$C_6H_5CH{=}CH{-}HgBr + C_6H_5ICl_2 \rightarrow C_6H_5CH{=}CH{-}\overset{\oplus}{I}C_6H_5\quad I^{\ominus}$$

nucleophiles is higher than that of the phenyl group in vinylphenyliodonium ions.[70] In the thermolysis of vinylphenyliodonium halides, vinyl halides and

$$CH_2{=}CH{-}\overset{\oplus}{I}(\overset{\ominus}{X}){-}C_6H_5 \rightarrow CH_2{=}CH{-}X + C_6H_5I$$

iodobenzene are formed. Aryliodonio-5,5-dimethylcyclohexane-1,3-dione, synthesized from aryliodoso acetate and 5,5-dimethylcyclohexane-1,3-dione, when reacted with acid, gave an iodonium ion bearing an olefinic group as a ligand.[71]

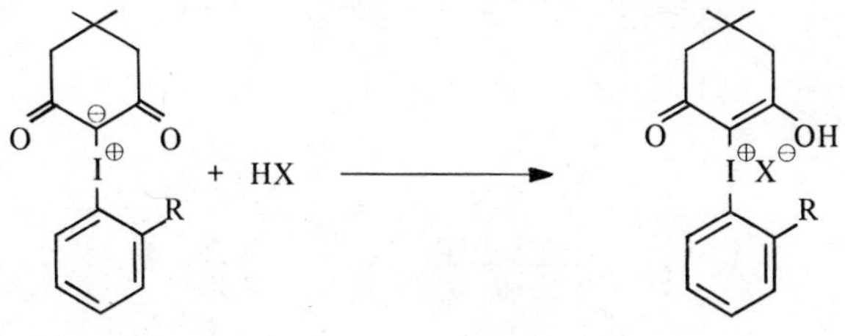

$$R = H,\ NO_2,\ CH_3$$

Phenyliodoso acetate reacts with phosphonium ylides to give the corresponding iodonium ions.[72] In the reaction of the iodonium ions (R = OEt) with

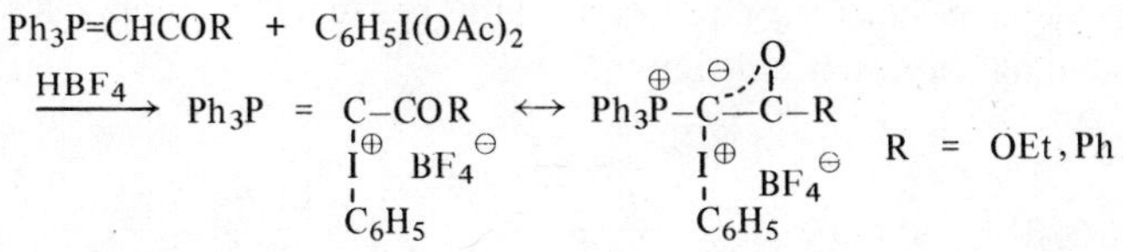

$$Ph_3P{=}CHCOR + C_6H_5I(OAc)_2 \xrightarrow{HBF_4} Ph_3P = C(\overset{\oplus}{I}C_6H_5)COR\ BF_4^{\ominus} \leftrightarrow Ph_3\overset{\oplus}{P}{-}\overset{\ominus}{C}(\overset{\oplus}{I}C_6H_5){-}C(=O){-}R\ BF_4^{\ominus} \qquad R = OEt, Ph$$

hydrogen bromide the C–I bond was cleaved to give the corresponding

$$Ph{-}\overset{\oplus}{I}{-}C(CO_2Et)(\overset{\oplus}{P}Ph_3)\ \overset{\ominus}{B}F_4 + HBr \xrightarrow[-Ph]{} Ph_3\overset{\oplus}{P}{-}CH(Br){-}CO_2Et\ BF_4^{\ominus}$$

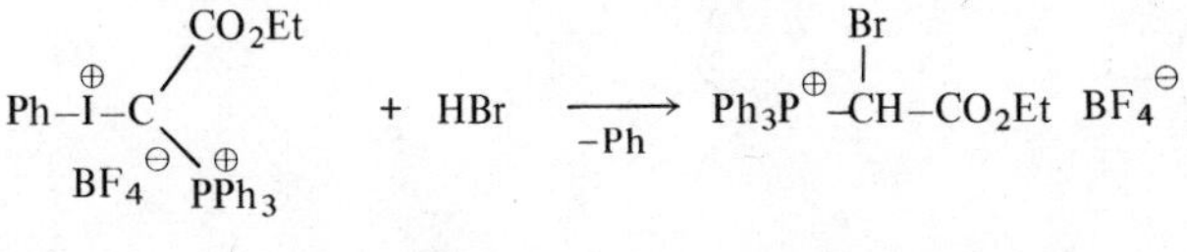

phosphonium ions. However, in the reaction of iodonium ions bearing a phenyl group with hydrogen chloride, phosphonium ylides were obtained.

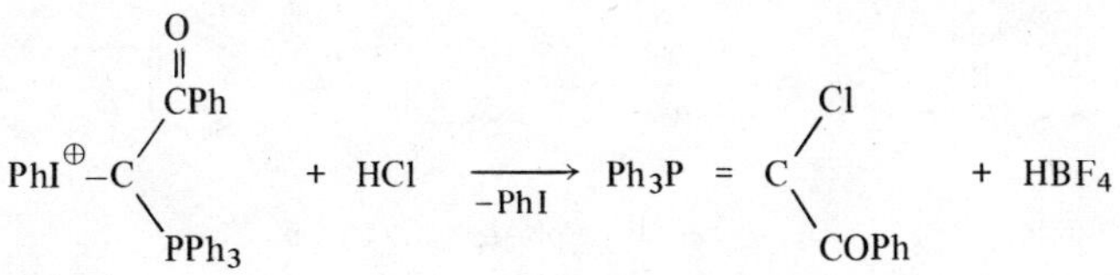

$$PhI^{\oplus}{-}C(\overset{O}{\overset{\|}{C}}Ph)(PPh_3) + HCl \xrightarrow[-PhI]{} Ph_3P = C(Cl)(COPh) + HBF_4$$

Di- and Polyiodonium Ions

The first synthesis of a diiodonium ion was that of the 2,5-thiophenediyl-bis(phenyliodonium) ion prepared from phenyl-2-thienyliodonium ions and aryliodoso acetate by Jezic.[73] This reaction has been proposed to be an electrophilic substitution reaction of aryliodoso acetate at the 4-position of the thienyl group of the iodonium ion.

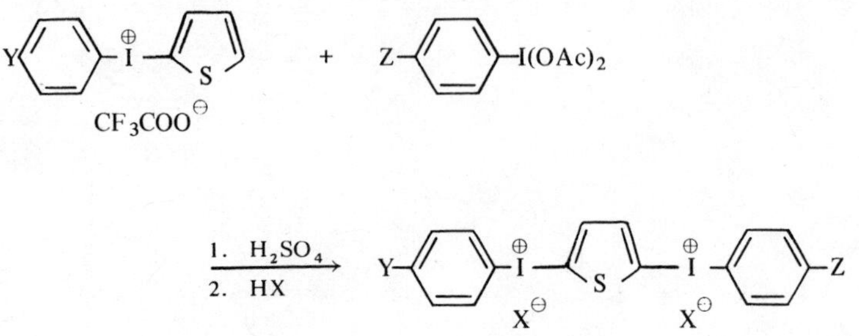

Y, $I^{\oplus}$, S, $CF_3COO^{\ominus}$ + Z, $I(OAc)_2$

1. H_2SO_4
2. HX

Y, $I^{\oplus}$, S, $I^{\oplus}$, Z, $X^{\ominus}$, $X^{\ominus}$

X = $HSO_4^{\ominus}$, $Cl^{\ominus}$, $Br^{\ominus}$, $I^{\ominus}$, $NO_3^{\ominus}$

Y = H, CH_3, Cl

Z = H, CH_3, Cl

p-Phenylenediiodoso acetate reacts with benzene or thiophene to give the corresponding bisiodonium ions.[74]

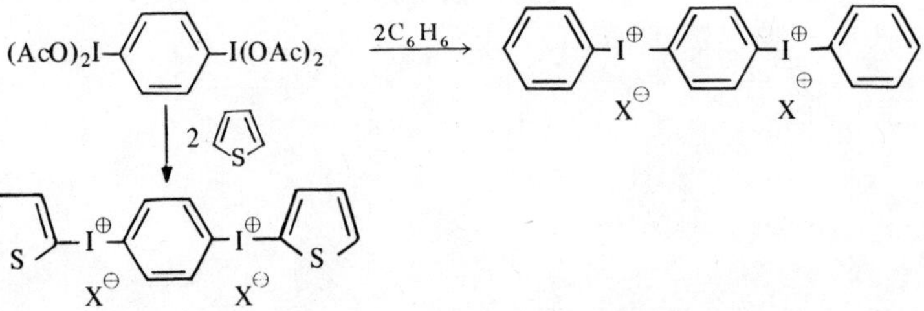

$(AcO)_2I$, $I(OAc)_2$

$2C_6H_6$

$I^{\oplus}$, $I^{\oplus}$, $X^{\ominus}$, $X^{\ominus}$

2 S

S, $I^{\oplus}$, $I^{\oplus}$, S, $X^{\ominus}$, $X^{\ominus}$

Substituted phenyliodoso acetates were treated in concentrated H_2SO_4 to give tetraiodonium ions in a self-condensation reaction of aryliodoso acetate.[74] These polyiodonium ions contain the cationic iodine in the

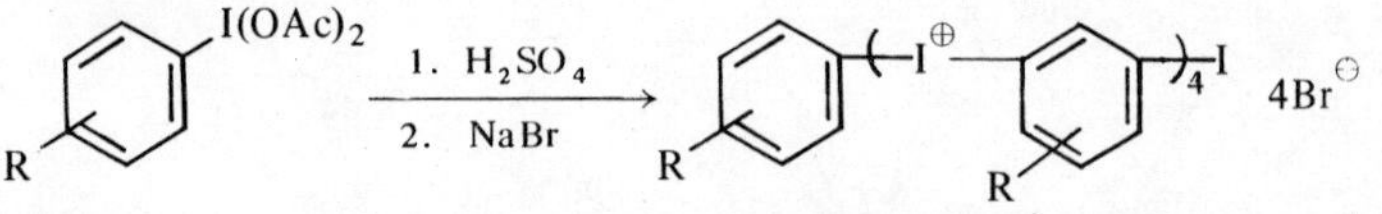

$I(OAc)_2$, R

1. H_2SO_4
2. NaBr

R, $I^{\oplus}$, R, 4, I, $4Br^{\ominus}$

R = *m*–Me, *o*–Me, *m*–Cl, *p*–Cl

main chain. Yamada and Okawava[75] synthesized a polyiodonium ion from polystyryliodoso acetate and benzene containing the cationic iodine in the side chain.

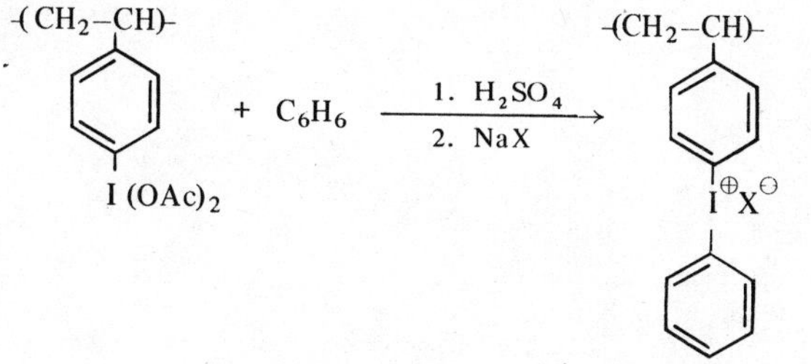

$$X^{\ominus} = HSO_4^{\ominus},\ Br^{\ominus},\ SCN^{\ominus},\ CF_3COO^{\ominus}$$

5.1.2 Kinetics of Formation

In the reaction of substituted phenyliodoso acetate with toluene, the reaction is first-order in both components under pseudo-first-order conditions.[76] H_2SO_4 has a very marked catalytic effect (a 30-fold change in acid concentration corresponds to a 5000-fold change in rate). The products are about 90% para isomer and 10% ortho isomer. Meta isomers are generally not formed. LeCount et al.[77] have also reported studies of the isomer distributions.

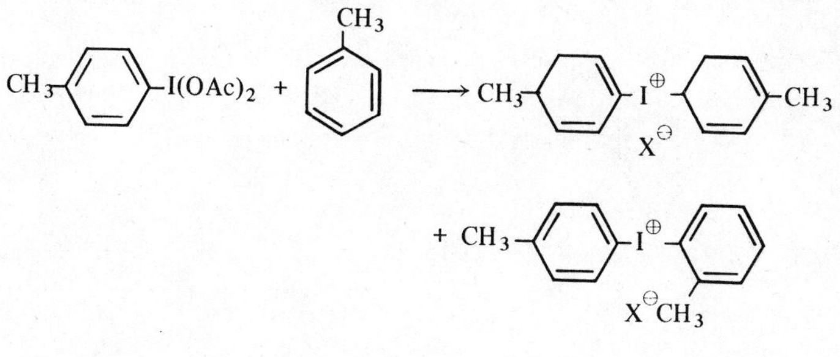

5.1.3 Structural Aspects

Diaryliodonium halides are generally ionic substances; the interaction between cationic iodine and the counteranions depends on the nature of the counteranions. Depending on the counterion, not only ionic, but also covalent character can be involved, such as in diaryliodonium chlorides.

X-Ray Studies

Medin[78] first showed by X-ray studies that in diphenyliodonium iodide the distance between the two iodine atoms is incompatible with a covalent link, and that the compound is a true iodonium salt.

Khotsyanova studied the crystal structure of diphenyliodonium chloride.[79] The molecule is T-shaped, containing a linear $C\text{-}1_{Ar}$–I–Cl group perpendicular to the other $C\text{-}2_{Ar}$–I group as shown. If the two nonbonded pairs of electrons

98° C_2 87°

C_1—I—Cl

175°

$C_1–I = C_2–I = 2.08Å$

$I–Cl = 3.08Å$

on iodine are taken into account, the molecule should have a trigonal bipyramidal structure.

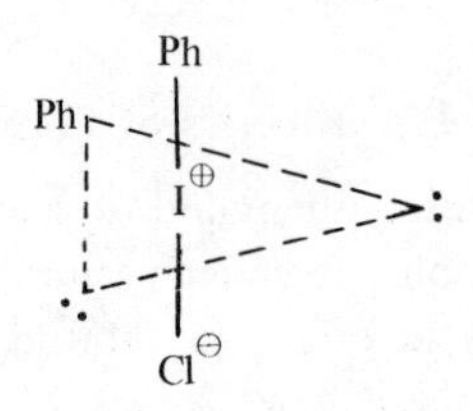

Ir Spectra

Many substituted diphenyliodonium ions were investigated by ir spectroscopy. In the case of diphenyliodonium iodide, the following major absorptions were found.[80] (The corresponding data for iodobenzene are shown in paren-

C–C stretching	1563 cm^{-1}	(1473 cm^{-1})
Ring breathing	992 cm^{-1}	(987 cm^{-1})
Out-of-plane C–H	742 cm^{-1}	736 cm^{-1})

thesis.) Arshad et al.[81] concluded the presence of interaction of the non-bonded pairs of electrons with the ring π electrons from the appearance of the weak absorption at 1590 cm^{-1}.

Uv Spectra

The uv spectra of a series of substituted diphenyliodonium ion have been studied[82] (Table 21). The wavelength of B bands shifts to the longer wavelength when the substituent group is changed to a strongly electron-withdrawing group. However, in the meta isomer such a shift is not observed. This indicates the presence of $p\pi$-$d\pi$ resonance between the phenyl ring and the cationic iodine center as shown. In the uv spectra of diaryliodonium iodide in ethanol,

TABLE 21
Uv Absorption Maxima of Diphenyliodonium Ions

Iodonium ion	Wavelength (nm)[a]	
	A band	B band
(C_6H_5-I)	207 (3.86)	226 (4.12)
$C_6H_5-\overset{\oplus}{I}-C_6H_5$	209 (4.23)	226 (4.19)
$C_6H_5-\overset{\oplus}{I}-C_6H_4-OCH_3$ (p)	210 (4.28)	245 (4.16)
$C_6H_5-\overset{\oplus}{I}-C_6H_4-OH$ (p)	211 (4.32)	246 (4.12)
$C_6H_5-\overset{\oplus}{I}-C_6H_4-NH_2$ (p)	211 (3.79)	285 (4.03)
$C_6H_5-\overset{\oplus}{I}-C_6H_4-O^{\ominus}$ (p)	222 (4.19)	292 (4.21)
$C_6H_5-\overset{\oplus}{I}-C_6H_4-OCH_3$ (m)	210 (4.39)	218 (4.38)

[a] Log ϵ is given in parentheses.

$$C_6H_5-\overset{\oplus}{I}-C_6H_4-OH \longleftrightarrow C_6H_5-I=C_6H_4=\overset{\oplus}{O}H$$

a shoulder absorption appears at 300-350 nm.[83] In methylene chloride, however, a new absorption maximum appears at 340-360 nm, which undergoes a bathochromic shift in passing to solvents with lesser polarity, along with increasing intensity of the band. Thus the iodide anions form a charge transfer complex with diaryliodonium ions. When the counteranion is tetrafluoroborate, chloride, or bromide, this is not observed.

Nmr Spectra

Based on pmr data for diphenyliodonium ions (Table 22), the π-resonance interaction between the $5d$ orbital of iodine and the phenyl ring has been

TABLE 22
Proton Nmr Shifts of Diphenyliodonium Ion Salts

Solvent	Anion	Aromatic proton		
		δ Ortho	δ Meta	δ Para
CH_2Cl_2	Cl^-	7.99	7.35	7.51
	NO_3^-	7.97	7.42	7.58
	AcO^-	7.91	7.33	7.48
CH_3OH	Cl^-	8.16	7.48	7.64
	NO_3^-	8.16	7.48	7.66
	AcO^-	8.15	7.48	7.66
D_2O	Cl^-	8.04	7.40	7.58
	NO_3^-	8.03	7.41	7.57
	AcO^-	8.06	7.42	7.58
Average shift difference		-0.63 ± 0.05	-0.16 ± 0.02	

discussed.[84] The differences between the ortho, para, and meta protons depend not only on the inductive effect of the positive iodine center, but also on the $d\pi$-$p\pi$ resonance between the cationic iodine and the phenyl rings. Petrosyan et al.[85] have also discussed the interaction between the diphenyliodonium ions and counteranions, based on nmr data (Table 23). In the case of the tetra-

TABLE 23
Proton Chemical Shifts of Diphenyliodonium Ion Salts in Dimethyl Sulfoxide

Counteranion	δ Ring protons (δ)[a]		
	Ortho	Meta	Para
BF_4^-	8.37 (8.17)	7.63	(7.62)
I^-	8.30	7.62	
Br^-	8.27	7.59	
Cl^-	8.22 (8.17)	7.51	(7.62)
CH_3COO^-	8.17 (8.16)	7.53	(7.61)

[a]Values obtained in methanol are given in parentheses.

fluoroborate anion, the iodonium ion is quite free. However, in the case of other more nucleophilic counteranions, increasing interaction is present between the iodonium ion and its counteranion. In methanol, the chemical shifts show no variation and are all similar to those of the BF_4^- salts. They suggest that no significant interaction between iodonium ions and counteranions exists in this solvent.

Nesmeyanov et al.[86] have calculated from the nmr data the effective charge at the para-carbon atom in onium compounds relative to their neutral molecular analogs (Table 24).

TABLE 24
Effective Charge at Para Carbons of Phenyl-Substituted Onium Ions Relative to Their Neutral Analogs

	$C_6H_5X^{\oplus}C_6H_5\ BF_4^{\ominus}$			$(C_6H_5)_3X^{\oplus}BF_4^{\ominus}$		
X	$I^{\oplus}$	$Br^{\oplus}$	$Cl^{\oplus}$	$S^{\oplus}$	$O^{\oplus}$	$C^{\oplus}$
Charge	0.027	0.047	0.063	0.053	0.086	0.106

Olah et al.[87] studied the carbon-13 spectra of diphenyliodonium and -bromonium ions in SO_2 solution at -40° (Table 25). Charge delocalization into the phenyl rings indicates the contriubtion of resonance forms of a quinoidal structure.

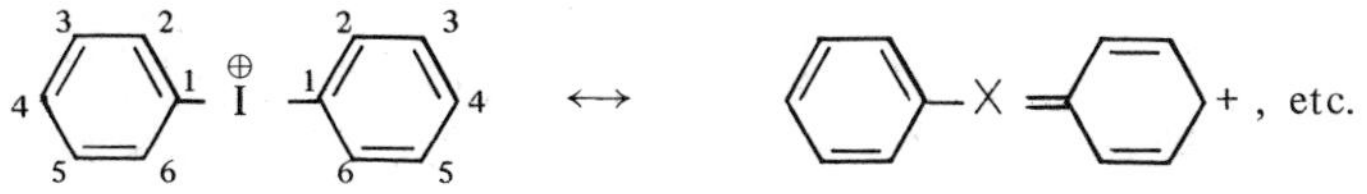

X = I, Br

TABLE 25
Carbon-13 Chemical Shifts of Diarylhalonium Ions[a]

$X^{\oplus}$	C-1	C-2	C-3	C-4	C-5	C-6
$Br^{\oplus}$	133.3	131.1	134.4	135.0	134.4	141.1
$I^{\oplus}$	111.1	131.5	128.7	129.2	128.7	131.5

[a] In parts per million from TMS.

Substituent Effect of the Phenyliodonio Group

Beringer et al.[88] determined the acidity of various benzoic acids and phenols bearing a phenyliodonio group (Table 26). The phenyliodoniobenzoic acids ($C_6H_5\overset{\oplus}{I}C_6H_4COOH$) are stronger acids than 4-nitrobenzoic acid ($pK_a = 5.83$ in CH_3CN-H_2O) (15:1). This is based on the electrostatic effect of the phenyliodonio group.

TABLE 26
pK_a Values of Acidic Diaryliodonium Ions

Acid	Apparent ionization constant, pK_a[a]
$C_6H_5-I^{\oplus}-C_6H_4$ (o-COOH)	3.5[b] (6.24)
$C_6H_5-I^{\oplus}-C_6H_4$ (m-COOH)	5.55[b] (6.45)
$C_6H_5-I^{\oplus}-C_6H_4$-COOH (p)	4.60[b] (6.47)
$C_6H_5-I^{\oplus}-C_6H_4$-OH (p)	8.33[c] (9.66)

[a]Values for the corresponding iodide compound are given in parentheses.
[b]In CH_3CN-H_2O (15:1).
[c]In EtOH-H_2O (3:7).

Degree of Dissociation

From conductivity data of diphenyliodonium salts at 5° in methanol and water, the degree of dissociation was calculated[89] (Table 27). In both water and methanol the degree of dissociation of diphenyliodonium fluoroborate is higher than that of other iodonium salts.

TABLE 27
Degree of Dissociation of Diphenyliodonium Salts

	Counterion			
Solvent	$BF_4^{\ominus}$	$Cl^{\ominus}$	$Br^{\ominus}$	$OAc^{\ominus}$
H_2O	97.8	90.8	88.2	86.3
CH_3OH	98.4	89.2	85.2	80.9

5.1.4 Chemical Reactivity

Reactions with Anionic Reagents

Diphenyliodonium iodide reacts with phenyllithium at −80° to give triphenyliodine which decomposes at −10° to biphenyl and iodobenzene.[90] The phenyla-

$$(C_6H_5)_2I^{\oplus}I^{\ominus} + C_6H_5Li \xrightarrow{-80^\circ} (C_6H_5)_3I + LiI \xrightarrow{-10^\circ} C_6H_5\text{-}C_6H_5 + C_6H_5I$$

tion of β-diketones, such as dimedone, and malonic esters by diphenyliodonium ion was studied.[91] It has been demonstrated, on the basis of products, that the reaction proceeds by a radical mechanism. Hauser et al.[92] have reported

$$ArI^{\oplus}Ar + R^{\ominus} \rightarrow \underset{\text{ion pairs}}{ArI^{\oplus}ArR^{\ominus}} \xrightarrow[-ArI]{\text{overall reaction}} ArR + Ar\text{-}Ar + R\text{-}R$$

$$\underset{\text{radical pairs}}{Ar\dot{I}Ar\ \dot{R}} \xrightarrow[-ArI]{} \underset{\text{radical pairs}}{Ar\cdot R\cdot} \rightleftharpoons \underset{\text{free radical}}{Ar\cdot + R\cdot}$$

many examples of the phenylation of active methylene groups by using diphenyliodonium ions. However, in an attempted phenylation of 2,4,6-

$$CH_3\text{-}\overset{O}{\overset{\|}{C}}\text{-}CH_2\text{-}\overset{O}{\overset{\|}{C}}\text{-}CH_3 + NaNH_2 + C_6H_5\overset{\oplus}{I}C_6H_5 \rightarrow CH_3\text{-}\overset{O}{\overset{\|}{C}}\text{-}CH_2\text{-}\overset{O}{\overset{\|}{C}}\text{-}CH_2\text{-}Ph$$

trimethylacetophenone, no phenylated product was obtained; only polymer formation took place.[93]

$$2,4,6\text{-}(CH_3)_3C_6H_2COCH_3 \xrightarrow[(C_6H_5)_2\overset{\oplus}{I}Cl^{\ominus}]{t\text{-BuOK}} \left(\overset{C_6H_5}{\underset{C=O\text{-}C_6H_2(CH_3)_3}{C}}\right)_n$$

4,4′-Dichlorodiphenyliodonium chloride reacts with phenyllithium to give bis(4-chlorophenyl)phenyliodine which on cleavage with hydrogen chloride gives chlorobenzene and phenyl,p-chlorophenyliodonium chloride.[94]

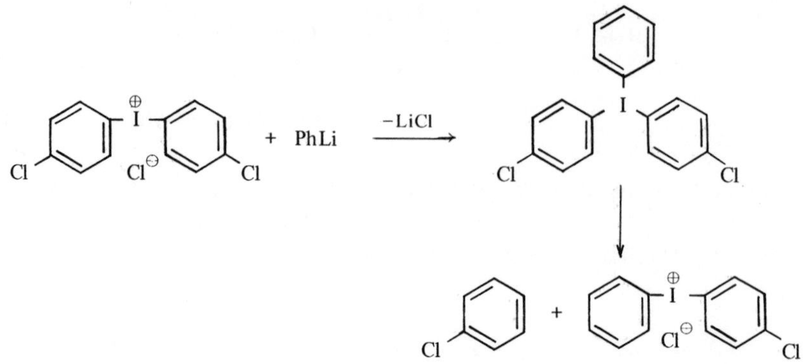

In the reaction of diphenyliodonium chloride with an excess of 4-dimethylaminophenyllithium, 4,4′-bis(dimethylamino)diphenyliodonium chloride was obtained (formed in a ligand exchange reaction). Diphenyliodonium chloride, when treated with 2,2′-dilithiobiphenyl, gave after carbonation and acidification dibenziodophenium ion and benzoic acid.

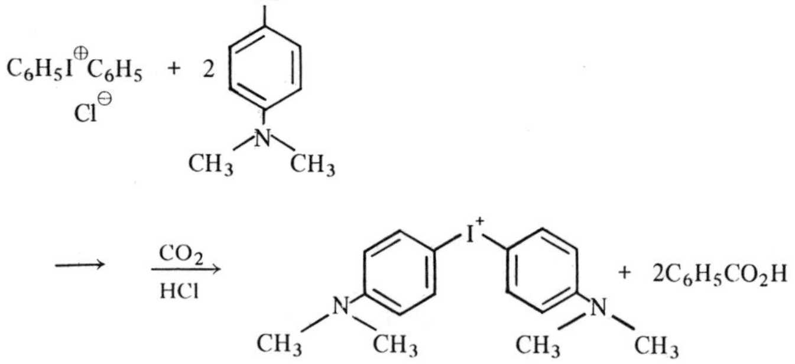

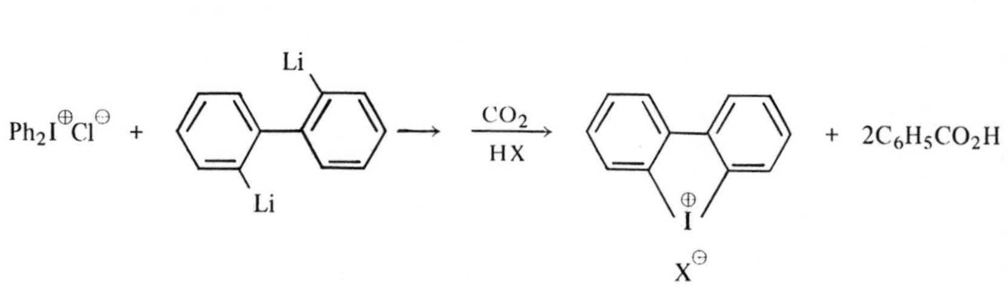

Active methylene groups are easily phenylated to give the corresponding phenyl compounds.[93]

$$\text{2-cyano-1-tetralone} \xrightarrow[Ph_2I^{\oplus}Cl^{\ominus}]{t\text{-BuOK}} \text{2-cyano-2-}C_6H_5\text{-1-tetralone}$$

In the reaction of isopropyl phenyl ketone with diphenyliodonium ion, on treatment with potassium isoamyloxide, the phenylated compound was obtained

$$C_6H_5-\overset{O}{\overset{\|}{C}}-CH(CH_3)_2 \xrightarrow[Ph_2I^{\oplus}Cl^{\ominus}]{\text{i-amylOK}} C_6H_5-\overset{O}{\overset{\|}{C}}-C(CH_3)_2-C_6H_5$$

in 57% yield. In the phenylation of 2-formyl-1-indanone, decarbonylation occurs and the diphenylated product is obtained in 74% yield.

$$\text{2-formyl-1-indanone} \xrightarrow[-CO]{t\text{-BuO}^{\ominus}} [\text{indenolate } O^{\ominus}] \xrightarrow[t\text{-BuO}^{\oplus}]{(C_6H_5)_2I^{\ominus}} [\text{2-}C_6H_5\text{-indenolate } O^{\ominus}] \xrightarrow{(C_6H_5)_2I^{\oplus}} \text{2,2-}(C_6H_5)_2\text{-1-indanone}$$

Diphenyliodonium ions react with tritylmagnesium chloride in tetrahydrofuran to give diphenyliodine radicals which were directly observed by esr spectroscopy.[45]

$$Ph_2I^{\oplus}I^{\ominus}\ (Cl^{\ominus},\ Br^{\ominus}) + Ph_3CMgCl \rightarrow [C_6H_5\dot{I}C_6H_5]$$

In the reaction with methylmagnesium iodide, methane and ethane are formed. The formation of ethane indicates that the methyl radical is an intermediate, which in turn must also mean that the diphenyliodine radical is an intermediate.

$$Ph_2I^{\oplus} + CH_3^{\ominus} \rightarrow [Ph\text{-}\dot{I}Ph] + CH_3\cdot$$

$$[Ph\text{-}\dot{I}Ph] \downarrow \text{decomposition} \qquad CH_3\cdot \downarrow CH_3-CH_3$$

Phenyl(β-phenylethynyl)iodonium chloride reacts with the carbanion of 2-phenyl-1,3-indandione to give 2-phenyl-2-phenylethynyl-1,3-indandione in 73% yield.[46] Likewise, in the reaction with phenyl(α-chlorostyryl)iodonium

$$C_6H_5-C\equiv C-\overset{\oplus}{I}-C_6H_5 \quad Cl^{\ominus} \quad + \quad \text{(2-phenyl-1,3-indandione anion, } O^{\ominus}\text{)} \longrightarrow \text{2-}C_6H_5\text{-2-}(C\equiv C-C_6H_5)\text{-1,3-indandione}$$

fluoroborate, 2-phenyl-2-(α-chlorostyryl)-1,3-indandione was obtained in 67% yield by cleavage of the bond between iodine and the vinyl group.

$$(C_6H_5)(Cl)C=C(H)-\overset{\oplus}{I}-C_6H_5 \quad BF_4^{\ominus} \quad + \quad \text{(2-phenyl-1,3-indandione anion, } O^{\ominus}\text{)} \longrightarrow \text{2-}C_6H_5\text{-2-}[CH=C(C_6H_5)Cl]\text{-1,3-indandione}$$

Reactions with Amines

Aromatic Amines. The reaction of diphenyliodonium ion with aniline is not a conventional S_N2 displacement.[97] Diphenyliodonium fluoroborate and aniline form a relatively stable complex (a charge transfer complex absorption appears at 320 nm). The use of an excess of aniline increases, whereas an excess of iodonium salt decreases, the yield of diphenylamine formed.

$$Ph_2I^{\oplus}BF_4^{\ominus} + C_6H_5NH_2 \xrightarrow[\text{i-PrOH}]{} PhI + PhNHPh + HBF_4$$

From the results, the following ionic reaction mechanism was proposed.

$$Ph_2I^{\oplus}BF_4^{\ominus} + PhNH_2 \rightleftharpoons [Ph_2I\cdot PhNH_2]^{\oplus}BF_4^{\ominus}$$

$$[C_6H_5-\overset{C_6H_5}{I}NH_2Ph]^{\oplus} BF_4^{\ominus} + Ph\ddot{N}H_2$$

$$\xrightarrow[-C_6H_5I]{} Ph_2NH + [PhNH_3]^{\oplus}BF_4^{\ominus}$$

As in the reaction not only diphenylamine but also benzene, biphenyl, and acetone (from isopropyl alcohol used as the solvent) were formed,[98] a different radical mechanism was proposed, accounting also for the side reactions.

$$(Ph_2I\cdot PhNH_2]^{\oplus}BF_4^{\ominus} \rightarrow Ph_2I\cdot + \cdot\overset{\oplus}{N}H_2Ph + BF_4^{\ominus}$$

$$Ph_2I\cdot \rightarrow Ph\cdot + PhI$$

$$Ph\cdot + H\cdot \text{ (from solvent)} \rightarrow Ph-H$$

$$2Ph\cdot \rightarrow Ph-Ph$$

$$\cdot NH_2Ph^{\oplus} + H\cdot \rightarrow PhNH_3^{\oplus}$$

$$(CH_3)_2CH-OH \xrightarrow{-2H\cdot} CH_3-\overset{\overset{O}{\parallel}}{C}-CH_3$$

The homolytic nature of the reaction was also indicated by the substituent effects.[99] However, when the counterion is halide ion (e.g., $Cl^{\ominus}$, $Br^{\ominus}$), not only diphenylamine but also halobenzene was formed.[100] Thus the following competing ionic reaction also occurs. Diphenyliodonium iodide is easily reduced

$$[C_6H_5-\overset{C_6H_5}{I}-NH_2Ph]^{\oplus} X^{\ominus} \xrightarrow[-C_6H_5\text{-}I]{} C_6H_5X + PhNH_2$$

by bis(N,N-dimethyl)phenylenediamine at room temperature to give the diphenyliodine radical and the radical cation of p-phenylenediamine, as shown by esr spectroscopy.[101] In the reaction of unsymmetric diaryliodonium ions

$$Ph_2I^{\oplus}I^{\ominus} + Me_2N-C_6H_4-NMe_2 \xrightarrow[\text{acetone}]{} Ph_2I\cdot + Me_2N-(C_6H_4)^{+\cdot}-NMe_2$$

$$Ph_2I\cdot \rightarrow Ph\cdot + PhI$$

$$Ph\cdot \rightarrow PhH$$

with aromatic amines, the more electropositive aryl group reacts with the amine to give the corresponding diarylamine.[102] Diphenyliodonium-2-carboxyl-

$$o\text{-}NO_2C_6H_4\overset{\oplus}{I}C_6H_5\ Br^{\ominus} + C_6H_5NH_2 \xrightarrow[H_2O]{\text{reflux}} o\text{-}NO_2C_6H_4\text{–}NH\text{–}C_6H_5$$

$$m\text{-}NO_2C_6H_4\overset{\oplus}{I}C_6H_5\ Br^{\ominus} + C_6H_5NH_2 \longrightarrow m\text{-}NO_2C_6H_4NHC_6H_5 + (C_6H_5)_2NH$$

ate reacts with 2,3-dimethylaniline in isopropyl alcohol to give the corresponding diarylamine.[103]

$$C_6H_5\overset{\oplus}{I}C_6H_4\text{-}2\text{-}COO^{\ominus} + 2,3\text{-}(CH_3)_2C_6H_3NH_2 \xrightarrow[\text{PrOH, reflux}]{Cu(OAc)_2} 2\text{-}HOOCC_6H_4\text{–}NH\text{–}C_6H_3\text{-}2,3\text{-}(CH_3)_2$$

Aliphatic Amines. Beringer et al.[104] have reported the reaction of *o*- and *m*-nitrodiphenyliodonium tetrafluoroborates with dimethylamine to give the corresponding arylated amines in 83 and 43% yield, respectively. However,

$$o\text{-}NO_2C_6H_4\overset{\oplus}{I}C_6H_5\ BF_4^{\ominus} + Me_2NH \xrightarrow{H_2O} o\text{-}NO_2C_6H_4NMe_2$$

$$m\text{-}NO_2C_6H_4\overset{\oplus}{I}C_6H_5\ BF_4^{\ominus} + Me_2NH \xrightarrow{H_2O} m\text{-}NO_2C_6H_4NMe_2$$

Reutov et al.[105] have reported the reaction of di-(*m*-nitrophenyl)iodonium ions with diethylamine to give only 4% a yield of *N,N*-diethyl-*m*-nitroaniline but a 100% yield of diethylammonium ions. In the reaction of the same iodonium ion with *n*-hexylamine, 9% *n*-hexyl-*m*-nitrophenylamine and 100% *n*-hexylammonium ion were obtained.

$$(m\text{-}NO_2C_6H_4)_2I^{\oplus}Br^{\ominus} + Et_2NH \xrightarrow{CH_3OH} m\text{-}NO_2C_6H_4\text{-}NEt_2 + Et_2\overset{\oplus}{N}H_2Br^{\ominus}$$

In the reaction of diphenyliodonium ions with diethylamine, only a trace of *N,N*-diethylaniline and 69% diethylammonium ion were obtained in acetone solution.[106] Diphenyliodonium ions react with triethylamine to give triethylammonium ions as the major product.

$$Ph_2I^{\oplus}BF_4^{\ominus} + NEt_3 \rightarrow H\overset{\oplus}{N}Et_3BF_4^{\ominus} + C_6H_6 + Ph\text{-}Ph + C_6H_5I$$

Diphenyliodonium ions also react with triethylamine in the presence of mercury to give diphenylmercury, although only in low yield.[107]

$$Ph_2I^{\oplus}BF_4^{\oplus} + Et_3N + Hg \xrightarrow[C_6H_6]{reflux} Ph\text{-}Hg\text{-}Ph$$

These results indicate formation of the diphenyliodine radical and the cation radical of triethylamine. Therefore the reaction of diaryliodonium ions with aliphatic amines is not a simple phenylation of an amine, but the following homolytic reaction sequence occurs.

$$Ar_2I^{\oplus}X^{\ominus} + NR_3 \rightleftharpoons [Ar_2I\cdot NR_3]^{\oplus}X^{\ominus}$$

$$[Ar_2I\cdot NR_3]^{\oplus}X^{\ominus} \rightarrow Ar_2I\cdot + [\cdot NR_3]^{\oplus}X^{\ominus}$$

$$2\ Ar_2I\cdot \rightarrow 2\ ArI + Ar\cdot$$

$$2Ar\cdot \rightarrow Ar\text{-}Ar$$

$$[\cdot NR_3]^{\oplus}X^{\ominus} \xrightarrow[\cdot H]{} [HNR_3]^{\oplus}X^{\ominus}$$

Reactions With Aziridines

In the reaction with aziridines, diphenyliodonium iodide acts as catalyst to give the corresponding olefin and oxazole[108] (Table 28). In the reaction,

$$X{-}C_6H_4{-}CO{-}\underset{\text{aziridine, }N{-}R}{CH{-}C(C_6H_5)H} \xrightarrow[\text{THF, reflux}]{Ph_2I^{\oplus}I^{\ominus}} X{-}C_6H_4{-}CO{-}CH{=}CH{-}C_6H_5 + X{-}C_6H_4{-}(\text{oxazole}){-}C_6H_5$$

TABLE 28
Products of the Reaction of Diphenyliodonium Ions with Substituted Aziridines

		Yield (%)	
X	R	*trans*-Benzalacetophenone	Oxazole
H	$CH_2C_6H_5$ (trans)	73	7
H	$CH_2C_6H_5$ (cis)	75	8
H	C_6H_{11} (trans)	76	
H	C_6H_{11} (cis)	72	
CH_3	$CH_2C_6H_5$ (trans)	52	44
CH_3	$CH_2C_6H_5$ (cis)	54	41
CH_3	C_6H_{11} (trans)	67	
CH_3	C_6H_{11} (cis)	65	

the diphenyliodonium ion attacks the carbonyl group, causing cleavage of the C–N bond.

$$Ar{-}C({=}O{\rightarrow}\overset{\oplus}{I}Ph_2){-}\underset{N{-}CH_2R'\ \leftarrow I^{\ominus}}{CH{-}CH(Ph)} \longrightarrow Ar{-}C(OIPh_2){=}C(H){-}CH(Ph){-}N{=}CHR' \equiv$$

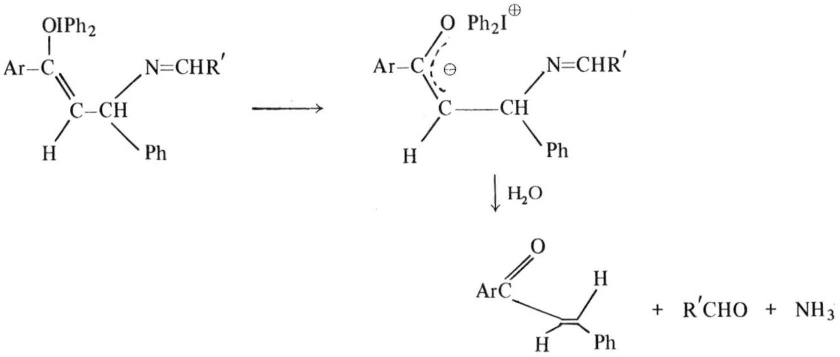

OIPh$_2$ | Ar–C | C–CH | N=CHR′ | H | Ph → O Ph$_2$I$^{\oplus}$ | Ar–C ⊖ | N=CHR′ | C——CH | H | Ph

↓ H_2O

ArC(=O) | H | H | Ph + R′CHO + NH$_3$

The oxazole is formed via C–C bond fission.

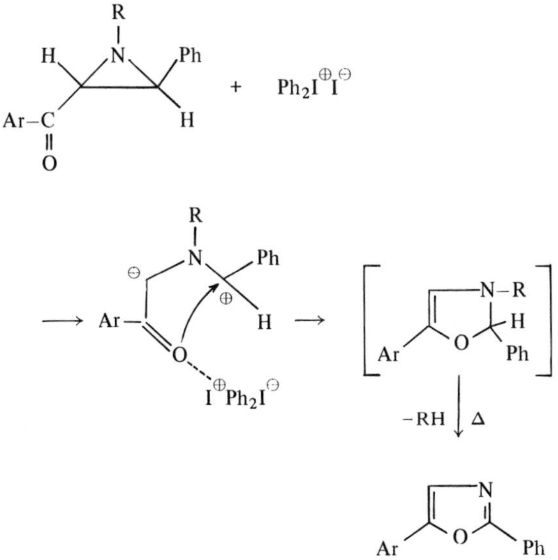

R | H | N | Ph | Ar–C | H | O + $Ph_2I^{\oplus}I^{\ominus}$

R | N | Ph | ⊖ | ⊕ | → Ar– | H | O | $I^{\oplus}Ph_2I^{\ominus}$ → [N–R | H | Ar | O | Ph]

–RH ↓ Δ

N | Ar | O | Ph

Reactions with Phosphines

Reactions with Triphenylphosphine. In the photolytic reaction of diphenyliodonium fluoroborate with triphenylphosphine, tetraphenylphosphonium fluoroborate was obtained in good yield.[109]

$$Ph_3P + Ph_2I^{\oplus}BF_4^{\ominus} \rightarrow Ph_4P^{\oplus}BF_4^{\ominus} + PhI$$

The presence of hydroquinone lowers the yield of phosphonium salt and, when ethanol was used as the solvent, benzene and acetaldehyde were formed as by-products. A radical mechanism was proposed, and intermediate formation of $[Ph_2\overset{+}{I}PPh_3]^{\oplus}BF_4^{\ominus}$ (absorption maximum appears at 336 nm) was indicated.

$$Ph_2I^{\oplus}BF_4^{\ominus} + Ph_3P \rightarrow [Ph_2I{:}PPh_3]^{\oplus}BF_4^{\ominus}$$

$$[Ph_2I{:}PPh_3]^{\oplus}BF_4^{\ominus} \xrightarrow{h\nu} Ph_2I\cdot + Ph_3P\overset{\oplus}{\cdot} + BF_4^{\ominus}$$

$$Ph_2I\cdot + Ph_3P \rightarrow Ph_4P\cdot + PhI$$

$$Ph_4P\cdot + Ph_2I^{\oplus}BF_4^{\ominus} \rightarrow Ph_4P^{\oplus}BF_4^{\ominus} + Ph_2I\cdot$$

In the photolytic reaction of unsymmetric iodonium ions with triphenylphosphine, phosphonium ions were obtained in a similar manner.[110]

$$(m\text{-}NO_2C_6H_4)\overset{\oplus}{I}(C_6H_5)\ BF_4^{\ominus} + Ph_3P \xrightarrow{h\nu} [m\text{-}NO_2C_6H_4\text{-}PPh_3]^{\oplus}\ BF_4^{\ominus}$$

In the related thermal reaction, phosphonium ions are also formed.[111]

$$Ph_3P + Ph_2I^{\oplus}BF_4^{\ominus} \xrightarrow[\text{dark, PrOH}]{\text{reflux}} Ph_4P^{\oplus}BF_4^{\ominus} + PhI$$

The presence of hydroquinone decreases the yield of phosphonium salt. Thus the thermal reaction also proceeds by the radical mechanism.

Reactions with Alkyl Phosphites and Alkyl Phosphates

Diphenyliodonium iodide reacts with triethyl phosphite to give *p*-diiodobenzene and diethyl phenylphosphonate.[112]

$$(C_6H_5)_2I^{\oplus}I^{\ominus} + (EtO)_3P \rightarrow (EtO)_2P(C_6H_5){\rightarrow}O + I\text{-}C_6H_4\text{-}I$$

Apparently, two distinct reactions take place. In the first the nucleophilic attack by phosphorus of the triethyl phosphite occurs at the C-1 phenyl carbon of the diphenyliodonium ion to give a phosphonium ion which decomposes to give diethyl phenylphosphonate. In the second reaction nucleophilic attack

$$(EtO)_3P: + C_6H_5I^{\oplus}C_6H_5\ I^{\ominus} \xrightarrow{-C_6H_5I} (EtO)_2\overset{\oplus}{P}(C_6H_5){-}O{-}Et\ I^{\ominus} \xrightarrow{-EtI} (EtO)P(C_6H_5){\rightarrow}O$$

occurs at the C-4 position of the aromatic ring, leading to the formation of *p*-diiodobenzene. These reactions indicate the ambident nature and reactivity

$$C_6H_5\overset{\oplus}{I}C_6H_5 + P(OEt)_3 \longrightarrow C_6H_5I{=}C_6H_5(H)\overset{\oplus}{P}(OEt)_3$$

$$\xrightarrow{-H^{\oplus}} C_6H_5\overset{\oplus}{I}(\overset{\ominus}{I})C_6H_4\overset{\oplus}{P}(OEt)_3\ I^{\ominus}$$

$$\xrightarrow{I^{\ominus}} C_6H_5\overset{\oplus}{I}(\overset{\ominus}{I})C_6H_4I \qquad \xrightarrow{C_6H_5I} C_6H_5\overset{\oplus}{I}(\overset{\ominus}{I})C_6H_4\overset{\oplus}{I}(\overset{\ominus}{I})C_6H_5$$

$$\longrightarrow I{-}C_6H_4{-}I$$

of diaryliodonium ions. Diphenyliodonium fluoroborate reacts with trialkyl phosphate to give iodobenzene and phenol.[113]

$$Ph_2I^{\oplus}BF_4^{\ominus} + (RO)_3PO \rightarrow PhI + PhOH$$

Reaction with Mercury

Diphenyliodonium halides react with mercury to give the corresponding phenylmercuric halide and iodobenzene.[114] The reaction rate increases with

$$Ph_2I^{\oplus}X^{\ominus} + Hg \rightarrow Ph{-}HgX + PhI$$

reduced polarity of the solvent,[115] and the addition of a common ion effectively interferes with the reaction of mercury and diphenyliodonium halide.[116]

In the presence of iron or copper, diphenyliodonium fluoroborate reacts with mercury to give diphenylmercury.[117]

$$C_6H_5\overset{\oplus}{I}C_6H_5\ BF_4^{\ominus} + Hg \rightarrow C_6H_5{-}Hg{-}C_6H_5$$

Protic solvents such as methanol and water are preferred for the preperation of diphenylmercury. The effect of solvent on the reaction is shown in Table 29.

TABLE 29
Solvent Dependence of the Preparation of Diphenylmercury from Diphenyliodonium Ion and Mercury

Solvent	Yield (%)
CH_3OH	53
H_2O	27
Acetone	Trace
Dioxane	0
C_6H_6	0

Diarylmercuries are also obtained electrochemically from diaryliodonium hydroxides by using a mercury cathode[118] (Table 30).

$$(R{-}C_6H_4)_2I^{\oplus}OH^{\ominus} \xrightarrow[\text{Hg cathode}]{} R{-}C_6H_4{-}Hg{-}C_6H_4{-}R$$

O-*Phenylation Reactions*

The reaction of diphenyliodonium fluoroborate with methyl benzoate gives phenyl benzoate in low yield, formed by an ester exchange reaction.[119]

TABLE 30
Yield of the Electrochemical Preparation of Diphenylmercuries from Diphenyliodonium Ions $(RAr)_2I^+$

R	Yield (%)
H	51
CH_3	41
CH_3O	58

$$Ph_2\overset{\oplus}{I}\ BF_4^{\oplus} + C_6H_5COOMe \rightarrow PhI + MeOBz + PhOBz$$

In the reaction of diphenyliodonium fluoroborate with phosphonium ylide, *O*-phenylated phosphonium ions were obtained in 15% yield.[120]

$$Ph_3P^{\oplus} - {}^{\ominus}CH-\overset{\overset{O}{\|}}{C}-CH_3 + Ph_2I^{\oplus}\ BF_4^{\ominus} \longrightarrow [Ph_3\overset{\oplus}{P}-CH{=}\overset{\overset{O-Ph}{|}}{C}-CH_3]BF_4^{\ominus}$$

The reaction of di(4-methoxy-3,5-dimethylphenyl)iodonium iodide with *N*-acetyl-3,5-diiodo-*L*-phenylalanine ethyl ester proceeds in methanol (45°, 24 hr) in the presence of triethylamine and copper powder to give the arylated compound in 70% yield.[121] However, the reaction of di(4-methoxy-3,5-dimethyl-

NH–Ac, CH_2–CH–CO_2Et, I, I, OH + $(CH_3O$–(3,5-di-CH_3-phenyl)$)_2$–$I^{\oplus}$ $I^{\ominus}$ ⟶

CH_2–CH–CO_2Et, NHAc, CH_3, CH_3, O, CH_3, CH_3, CH_3

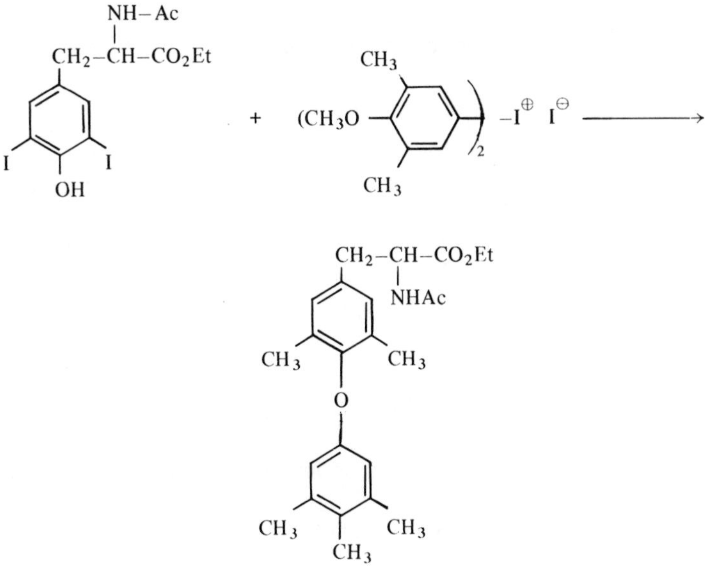

phenyl)iodonium trifluoroacetate with 4-bromo-2,6-dimethylphenol did not give the corresponding arylated compound; only poly-2,6-dimethylphenylene ether was obtained.[122]

Phenylation of oxime anions has also been reported. The reaction of diphenyliodonium bromide with the anion of benzophenone oxime gives the *O*-phenylated compound in 79% yield.[123] However, in the case of *p*-methylbenzophenone oxime not only *O*-phenylated but also *N*-phenylated products were obtained.

$$(C_6H_5)_2C{=}NO^{\ominus}Na^{\oplus} + Ph_2I^{\oplus}Br^{\ominus} \rightarrow (C_6H_5)_2C{=}N{-}OPh$$

$$(p\text{-}CH_3C_6H_4)(C_6H_5)C{=}N{-}O^{\ominus}Na^{\oplus} + Ph_2I^{\oplus}Br^{\ominus} \longrightarrow$$

$$(p\text{-}CH_3C_6H_4)(C_6H_5)C{=}N{-}OC_6H_5 + (p\text{-}CH_3C_6H_4)(C_6H_5)C{=}\overset{\oplus}{N}(O^{\ominus})C_6H_5$$

In the thermolytic reaction of diphenyliodonium hydroxide, diphenyl ether was obtained, which was formed in the reaction of diphenyliodonium ion and phenol, obtained from the thermolysis of diphenyliodonium hydroxide.[124]

$$Ph_2I^{\oplus}OH^{\ominus} \rightarrow PhOPh + PhOH + PhI + Ph{-}Ph$$

Reactions with Inorganic Nucleophiles

The reactions of diaryliodonium ions with inorganic nucleophiles such as nitrite ion, halide ions, and cyanide ion have been shown to give the corresponding aromatic compounds. The decreasing order of reactivity of various nucleophiles was found to be.[125]

$$Ar_2I^{\oplus} + X^{\ominus} \longrightarrow ArI + Ar{-}X$$

$$SO_3^{2-} \cong {}^{\ominus}OC_6H_5SO_3^{\ominus} >> NO_2^{\ominus} \cong CN^{\ominus} > Cl^{\ominus}$$

When cupric sulfate is present, cyanide ion is more reactive than nitrite ion.

$$SO_3^{2-} > CN^{\ominus} > NO_2^{\ominus}$$

In the reaction of unsymmetric iodonium ions with inorganic nucleophiles, there are two competing reaction paths (*a* and *b*). When bearing an electron-

$$\text{W-C}_6\text{H}_4\text{-}\overset{\oplus}{\text{I}}\text{-C}_6\text{H}_4\text{-R} + \text{X}^{\oplus} \xrightarrow{a} \text{W-C}_6\text{H}_4\text{-I} + \text{R-C}_6\text{H}_4\text{-Y}$$

$$\downarrow b$$

$$\text{W-C}_6\text{H}_4\text{-Y} + \text{R-C}_6\text{H}_4\text{-I}$$

W = Electron-withdrawing group

R = Electron-releasing group

withdrawing group the C-1 phenyl carbon is attacked more easily by the nucleophile than when bearing an electron-releasing group.[126] Beringer et al. have reported kinetic solvent effects in the reaction of diphenyliodonium ions with various nucleophiles.

The hydrolysis of diphenyliodonium ion proceeds homolytically.[127]

$$\text{Ar-}\overset{\oplus}{\text{I}}\text{-Ar} + 2H_2O \rightleftharpoons Ar_2IOH + H_2O^{\oplus}$$

$$Ar_2I\text{-}OH \longrightarrow ArI + Ar\cdot + \cdot OH$$

$$Ar\cdot + \cdot OH \longrightarrow ArOH$$

Copper(I) salts are extremely efficient for the decomposition of diaryliodonium ions.[128] In the decomposition of diphenyliodonium chloride, cuprous chloride acts as a good catalyst to give chlorobenzene, iodobenzene, and phenol. The effect of solvents on the reaction is shown in Table 31.

$$\text{C}_6\text{H}_5\text{-}\overset{\oplus}{\text{I}}\text{-C}_6\text{H}_5\ \text{Cl}^{\ominus} \xrightarrow{CuCl} C_6H_5Cl + C_6H_5OH + C_6H_5I$$

TABLE 31
Decomposition of Diphenyliodonium Ion by Cuprous Chloride

Solvent	C_6H_5Cl (%)	C_6H_5OH(%)
H_2O	95	5
CH_3OH	100	
Acetone	100	

The reaction proceeds by the following mechanism

$$Ph_2I^{\oplus}\ [CuCl_3]^{2-} \rightleftharpoons Ph_2I{-}CuCl_3^{\ominus} \rightarrow [Ph_2I \cdots CuCl_3^{\ominus}\ \text{(four-centre)}] \rightarrow PhI + PhCl + CuCl_2^{\ominus}$$

Chronium(V) and especially titanium(III) chlorides are also good catalysts in the decomposition of diphenyliodonium chloride in water to give quantitatively chlorobenzene and iodobenzene (Table 32).

$$(C_6H_5)_2I^{\oplus}\ Cl^{\ominus} \xrightarrow{CuCl} C_6H_5Cl + C_6H_6 + C_6H_5{-}C_6H_5 + C_6H_5I$$

TABLE 32
Decomposition of Diphenyliodonium Ion with Chromium(V) Chloride and Titanium(III) Chloride

Solvent	Catalyst	Reaction (%)	C_6H_5Cl	C_6H_5	$C_6H_5-C_6H_5$
H_2O	$TiCl_3$	100	100		
CH_3OH	$TiCl_3$	10	100		
H_2O	$CrCl_3$	61	11	50	39

The reaction of diphenyliodonium nitrate with carbon monoxide, carried out in methanol at 110° under 120 atm, has been reported to give methyl benzoate and benzoic acid.[129]

$$(C_6H_5)_2I^{\oplus}\ NO_3^{\ominus} + CO \xrightarrow[\text{in } CH_3OH]{} C_6H_5CO_2CH_3 + C_6H_5CO_2H + C_6H_5I + HNO_3$$

In the reaction of mesityl-*p*-tolyliodonium bisulfite in water, the nucleophile attacks the C-1 *p*-tolyl carbon more easily than the C-1 mesityl carbon.[131] This can be explained by the obvious steric hindrance of the two ortho methyl groups of the mesityl group. However, Yamada and Okawara[131] have

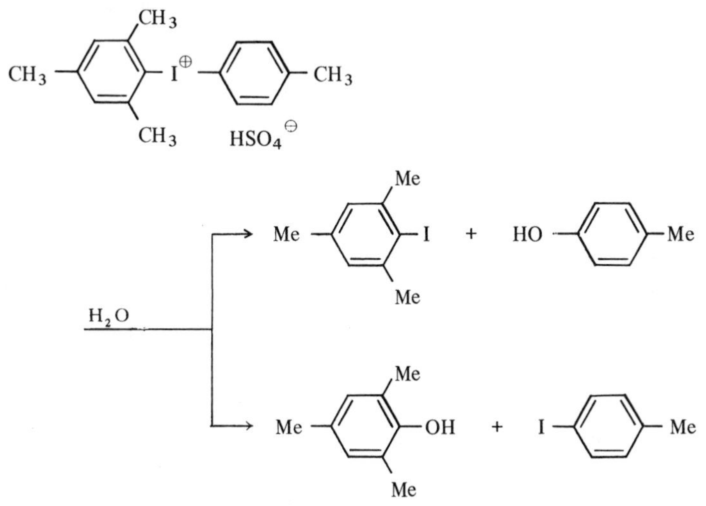

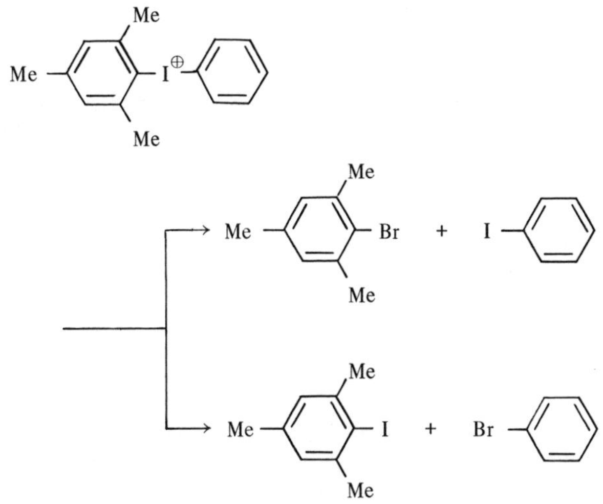

reported a reverse steric effect. In the reaction of *o*-tolymesityl- and phenylmesityliodonium bromide, bromide ion attacks the *o*-substituted C-1 phenyl carbon preferentially. These findings are best explained by the effects of intramolecular strain, due to the presence of ortho methyl groups, causing the reaction to proceed also by an S_N1 mechanism.

Reactions of Hetroaryliodonium Ions

The thermolytic reaction of phenyl-2-thienyliodonium iodide gives iodobenzene and 2-thienyl iodide.[57] Phenyl-2-thienyliodonium ion reacts with sulfite ion to give thiophene 2-sulfonate and iodobenzene, or benzenesulfonate and 2-iodothiophene. The ratio of attack on the thiophene ring to attack on the benzene ring is 2:1. The replacement of a phenyl group in the diphenyliodonium ion by a 2-thienyl group enhances halonium ion susceptibility to nucleophilic attack.

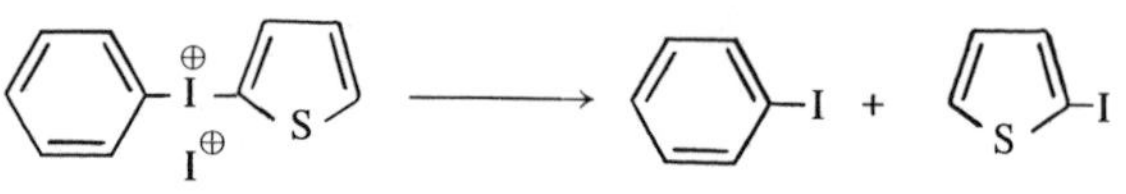

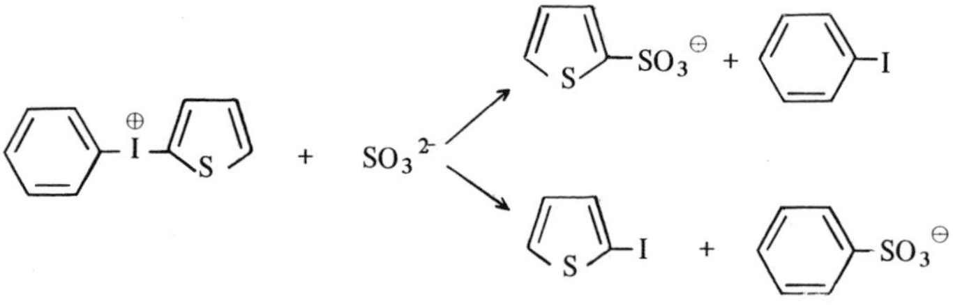

Yamada and Okawara[132] have, however, reported that chloride and bromide ions are more likely to attack the phenyl rather than the 2-thienyl ring in the reactions of substituted phenyl-2-thienyliodonium ions. Specifically, chloride ions attack the C-1 phenyl carbon, despite the presence of substituent groups.

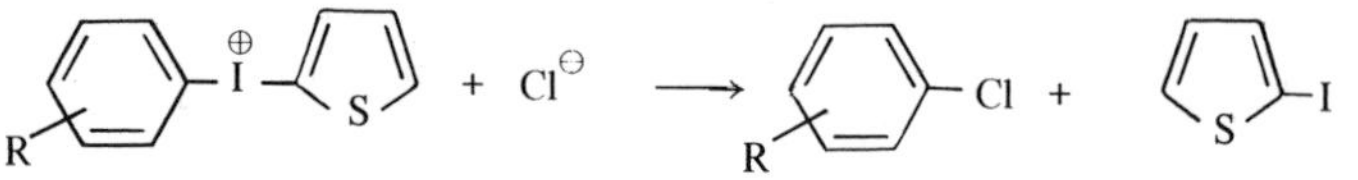

R = *m*-Me, *p*-Cl, *m*-Cl, H

Bromide ions attack not only the phenyl ring but also the thienyl ring to give the corresponding bromides. The phenyl ring is, however, much more reactive than the 2-thienyl ring (Table 33).

R–C₆H₄–I⊕–(2-thienyl) + Br⊕ —mode A→ R–C₆H₄–Br + 2-iodothiophene
R–C₆H₄–I⊕–(2-thienyl) + Br⊕ —mode B→ R–C₆H₄–I + 2-bromothiophene

TABLE 33
Mode of Attack of Bromide Ion on Phenyl-2-Thienyliodonium Ion in DMP at 100°

R	Mode A	Mode B
H	100	0
m-Cl	100	0
p-Cl	76.1	23.9
m-Me	100	0
p-Me	70.7	29.3
p-MeO	60.0	40.0

When methyl groups are present in the ortho position of the aryl-2-thienyl-iodonium ion, the aryl ring becomes much more reactive, as a result of obvious intramolecular strain.[131]

2,4,6-(CH_3)₃C₆H₂–I⊕–(2-thienyl) + X⊖ ⟶ 2,4,6-(CH_3)₃C₆H₂–X + 2-iodothiophene

5.1.5 Biological Activity

Positive iodine compounds have, in general, good oxidizing power, and as a result certain biocidal properties generally ascribed to "iodine"—tinctures, preparations, and complexes—were found for iodonium ions. Bis(3,4-dichlorophenyl)iodonium chloride and bis(2,4-dichlorophenyl)iodonium sulfate were evaluated as bactericides; the latter proved most effective against gram-negative bacteria.[136] Several substituted diaryliodonium chlorides were found to be inert to type-A influenza virus, although it was reported that ring chlorination increased the activity in this respect.[137,138] Diphenyliodonium salts were found to have active spermicidal properties in aqueous solutions of 0.5% or less.[139]

5.2 DIARYLBROMONIUM AND -CHLORONIUM IONS

Whereas diaryliodonium ions have been extensively studied, only very limited data are so far available on diarylbromonium and -chloronium ions.

Phenyldiazonium salts, as observed by Nesmeyanov,[133] react with bromo- and chlorobenzene to give the corresponding diphenylbromonium and -chloronium ions. The reaction, however, gives only exceedingly low yields. The decomposition of phenyldiazonium salts in halobenzene at 60° gives diphenylhalonium ion in 0.5% yield. If acetone is used as the solvent, the yield rises to 6%.

$$C_6H_5N_2^{\oplus}BF_4^{\ominus} + C_6H_5X \xrightarrow{60^\circ} C_6H_5X^{\oplus}C_6H_5\ BF_4^{\ominus}$$

$$\downarrow \text{in acetone}$$

$$C_6H_5X^{\oplus}C_6H_5\ BF_4^{\ominus} \qquad X = Cl,\ Br$$

Bromonium and chloronium ions are less stable than the corresponding iodonium ions and readily decompose to give halobenzenes. Diphenylbromonium and chloronium ions react with amines and pyridine to give the corresponding *N*-phenylated compounds. Halonium iodides react with mercury to give the corresponding phenylmercuric iodides.[133] However, when the counterion is the tetrafluoroborate anion, phenylmercuric compounds are not obtained.

$$Ph{-}X^{\oplus}{-}Ph\ I^{\ominus} + Hg \rightarrow Ph{-}Hg{-}I + PhX \qquad X = Cl,\ Br$$

In the photolysis reaction of 3,5-di-*tert*-butylbenzene-1,4-diazooxide with 2,6-diisopropyl-4-bromophenol, diarylbromonium bromide was obtained in 15%

O, NH2 (quinone diazide structure) + OH, Br (2,6-diisopropyl-4-bromophenol) $\xrightarrow{> 4800\ \text{Å}}$

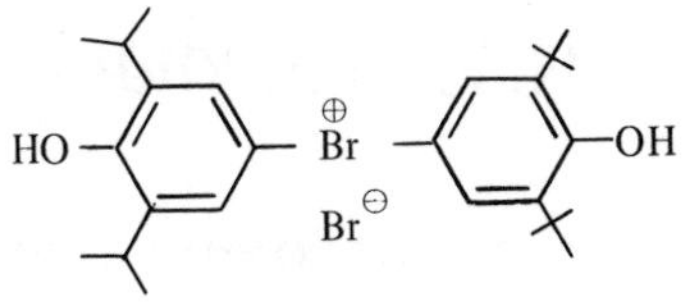

HO — Br⊕ — OH

Br⊖

yield.[134] A bromonium ylide was suggested to be the reaction intermediate.

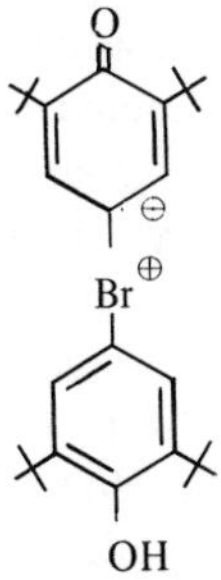

O

⊖

Br⊕

OH

CHAPTER 6

Halonium Ylides

6.1 ALKYLHALONIUM YLIDES

Dimethylhalonium ions, as discussed are good methylating reagents for a wide range of n and π donors.[3,26] The reaction with H_2SO_4 is typical.[24] When dimethylbromonium or iodonium ion complexes are added (as stable hexafluoroantimomate salts) to H_2SO_4, no reaction occurs below 0°, but at 30° dimethyl sulfate is formed. When D_2SO_4 is reacted at 30°, the pmr spectrum indicates that H-D exchange occurs during the course of the reaction. (When dimethylchloronium ion is reacted, it slowly dissolves in D_2SO_4 to give dimethyl sulfate, but no H-D exchange is observed.) After 2 hr, spectra of the H_2SO_4–soluble layer indicate the presence of dimethyl sulfate and a second species resonating at δ 4.00 (pmr) and δ 1.2 (cmr), which was shown to be the methylhydridoiodonium ion (CH_3I^+H). The H-D exchange indicates that methylenemethyliodonium and -bromonium ylides, the first reported alkylhalonium ylides, are intermediates.[24] Attempts to trap the ylides produced under basic conditions have so far been unsuccessful.

$$CH_3\overset{+}{X}CH_3\ SbF_6^- \xrightarrow{D_2SO_4} CH_3OSO_3H + CH_3X \xrightarrow{D_2SO_4} CH_3\overset{+}{X}D \xrightarrow{CH_3OSO_3H} (CH_3O)_2SO_2$$

$$CH_3\overset{+}{X}CH_3\ SbF_6^- \underset{}{\overset{HSO_4^-}{\rightleftharpoons}} [CH_3\overset{+}{X}\overset{-}{C}H_2] \underset{}{\overset{D_2SO_4}{\rightleftharpoons}} CH_3\overset{+}{X}CH_2D \rightleftharpoons \text{Further exchange}$$

X = Br, I

X = Br, I

6.2 ARYLHALONIUM YLIDES

Neilands and his coworkers[140-146] first investigated iodonium ylides. Phenyliodoso acetate reacts with active methylene groups in the presence of alkali to give the corresponding iodonium ylides.

$$C_6H_5I(OAc)_2 + CH_2\begin{matrix} R \\ R' \end{matrix} \xrightarrow[MeOH]{KOH} C_6H_5\overset{\oplus}{I}-\overset{\ominus}{C}\begin{matrix} R \\ R' \end{matrix} + 2AcOH$$

$$CH_2\begin{matrix} R \\ R' \end{matrix} = CH_2(CO_2CH_3)_2,^{140}$$ [indane-1,3-dione structure] 141

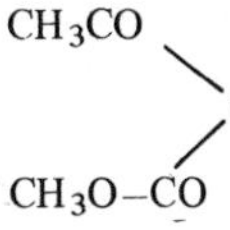

$CH_3CO(CH_3O{-}CO)CH_2$,[142] $(C_6H_5CO)_2CH_2$[143]

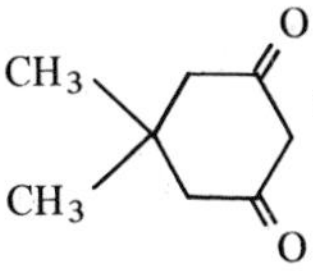

[144], 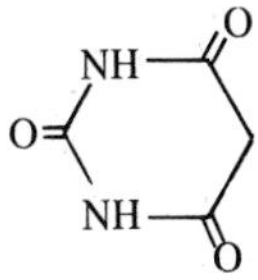 [145], 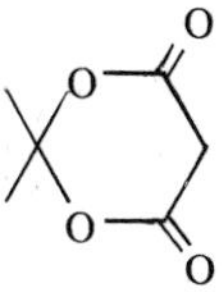 [146]

The iodonium ylides react with pyridine to give pyridinum ylides (by an ylide exchange reaction).[147]

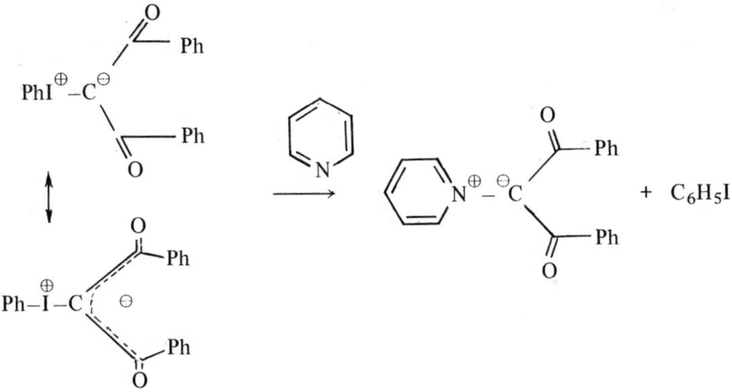

$$PhI^{\oplus}{-}C^{\ominus}(COPh)_2 \longleftrightarrow Ph{-}I^{\oplus}{-}C^{\ominus}(COPh)_2 \xrightarrow{C_5H_5N} C_5H_5N^{\oplus}{-}{}^{\ominus}C(COPh)_2 + C_6H_5I$$

In the reaction with acids, such as HCl and trichloroacetic acid, products of the addition of the acid to the carbanion center of the iodonium ylide were obtained, with cleavage of iodobenzene.

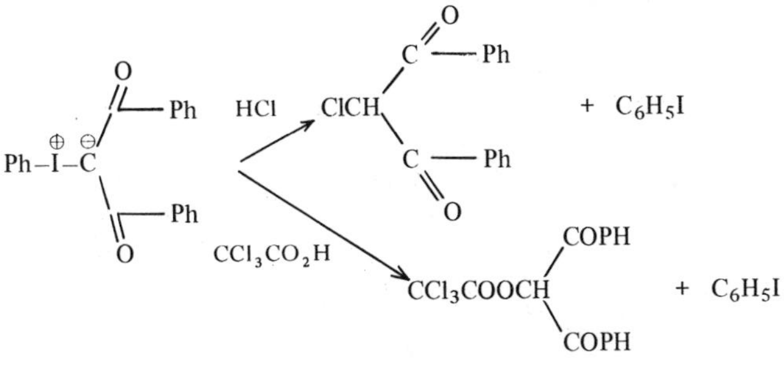

$$Ph{-}I^{\oplus}{-}C^{\ominus}(COPh)_2 \xrightarrow{HCl} ClCH(COPh)_2 + C_6H_5I$$

$$Ph{-}I^{\oplus}{-}C^{\ominus}(COPh)_2 \xrightarrow{CCl_3CO_2H} CCl_3COOCH(COPH)_2 + C_6H_5I$$

In the presence of cupric sulfate or cupric cyanide as a catalyst, the iodonium ylide reacts with pyridine to give pyridinium ylides in good yield.[147] In the reaction with the *p*-toluenesulfonic acid salt of *o*-aminobenzoic acid, the *N*-alkylated compound is obtained.[148] However, in the reaction with the

$$C_6H_5\overset{\oplus}{I}-\overset{\ominus}{C}(COPh)_2 + o\text{-}HOOC\text{-}C_6H_4\text{-}NH_3^{\oplus}\,Tos^{\ominus} \longrightarrow o\text{-}HOOC\text{-}C_6H_4\text{-}NHCH(COPh)_2 + C_6H_5I + TosOH$$

p-toluenesulfonic acid salt of α-amino acids, *O*-alkylated products are obtained.[149]

$$Ph\overset{\oplus}{I}-\overset{\ominus}{C}(COR)_2 + RR'C(COOH)(NH_3^{\oplus}\,Tos^{\ominus}) \xrightarrow{-PhI} RR'C(CO_2CH(COR)_2)(NH_3^{\oplus}\,Tos^{\ominus})$$

Iodonium ylides react with more acidic active methylene groups, transferring PhI to them to give new iodonium ylides and liberating weaker acids.

Hayashi et al.[150] have suggested that carbenes are intermediates in the decomposition of iodonium ylides. In the thermal reaction (in refluxing ethanol, 2 hr) dibenzoylmethane was obtained as a major product.

$$Ph-\overset{\oplus}{I}-\overset{\ominus}{C}(COPh)_2 \xrightarrow{\Delta} C_6H_5-\overset{O}{\overset{\|}{C}}-CH_2-\overset{O}{\overset{\|}{C}}-C_6H_5 + C_6H_5COCH(Ph)-CO_2Et + C_6H_5CO_2Et + C_6H_5I$$

In the presence of cuprous chloride the iodonium ylide decomposes spontaneously to give ethyl α-phenyl acetate, α-benzoyl acetate, and tetrabenzoylethylene as major products.

$$PhI^{\oplus}-C^{\ominus}(COPh)_2 \xrightarrow{CuCl} C_6H_5COCH(C_6H_5)CO_2Et + (C_6H_5CO)_2C{=}C(COC_6H_5)_2 + C_6H_5CO_2Et + C_6H_5I$$

The different distribution of the reaction products between thermal and catalytic decomposition was suggested to be dependent on the nature of the carbene intermediate. The following mechanism was suggested for the reactions.

$$Ph-I^{\oplus}-C^{\ominus}(COPh)_2 \xrightarrow[-PhI]{A} {\uparrow\downarrow}C(COPh)_2 \longrightarrow {\uparrow\downarrow}C(COPh)_2$$

$${\uparrow\downarrow}C(COPh)_2 \xrightarrow{EtOH} C_6H_5COCH{-}CO_2Et\ (C_6H_5) \qquad {\uparrow\downarrow}C(COPh)_2 \xrightarrow{EtOH} CH_2C(C(=O){-}Ph)_2$$

$$Ph-I^{\oplus}-C^{\ominus}(COPh)_2 \xrightarrow[r.t.]{CuCl} Cu\left[{\uparrow\downarrow}C(COPh)_2\right] \xrightarrow{EtOH} C_6H_5COCH{-}CO_2Et\ (C_6H_5)$$

$$Cu\left[{\uparrow\downarrow}C(COPh)_2\right] \longrightarrow (PhCO)_2C{=}C(COPh)_2$$

Diazodicyanoimidazole reacts with bromo- and chlorobenzene to give the corresponding bromonium and chloronium ylides which decompose to afford 1-phenyl-2-bromo-(or chloro)-4,5-dicyanoimidazole.[151]

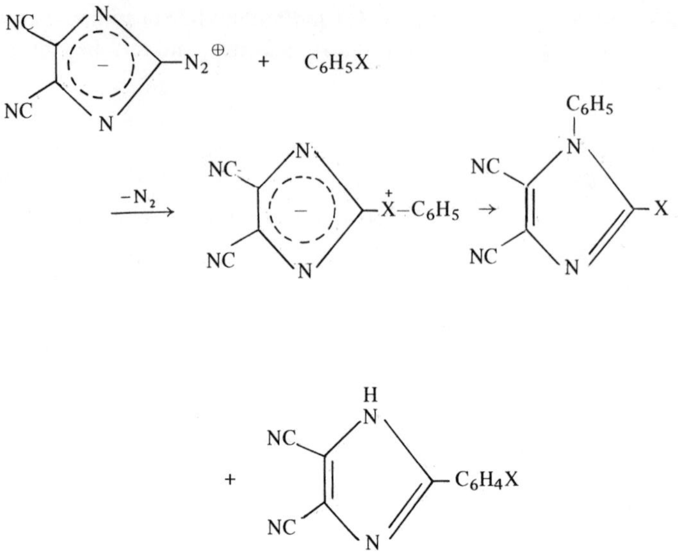

$X = Cl, Br$

Iodobenzene reacts similarly with diazodicyanoimidazole to give the dicyanoimidazole iodonium ylide which on decomposition gives 1-phenyl-2-iodio-4,5-dicyanoimidarole.[152]

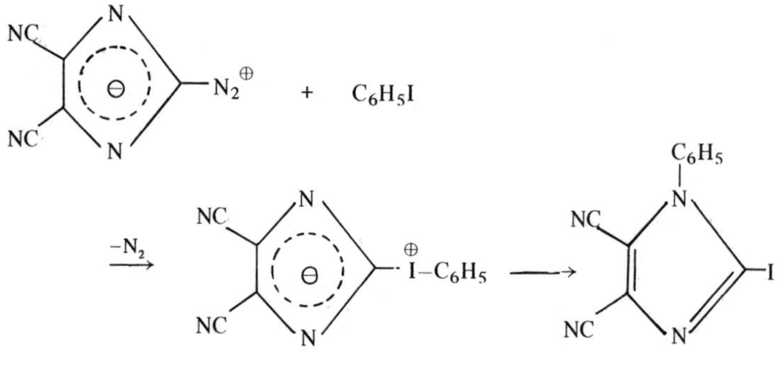

Yamada and Okawara[152] synthesized an I—N ylide from phenyliodoso acetate and *p*-toluenesulfonic acid in the presence of potassium hydroxide,

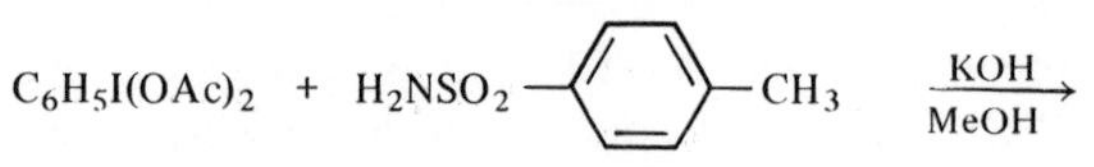

$$C_6H_5I(OAc)_2 + H_2NSO_2-C_6H_4-CH_3 \xrightarrow[MeOH]{KOH}$$

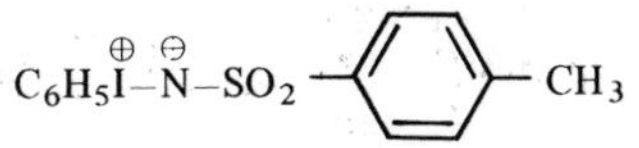

$$C_6H_5\overset{\oplus}{I}-\overset{\ominus}{N}-SO_2-C_6H_4-CH_3$$

which reacts with dimethyl sulfoxide to give a sulfoxylimine.

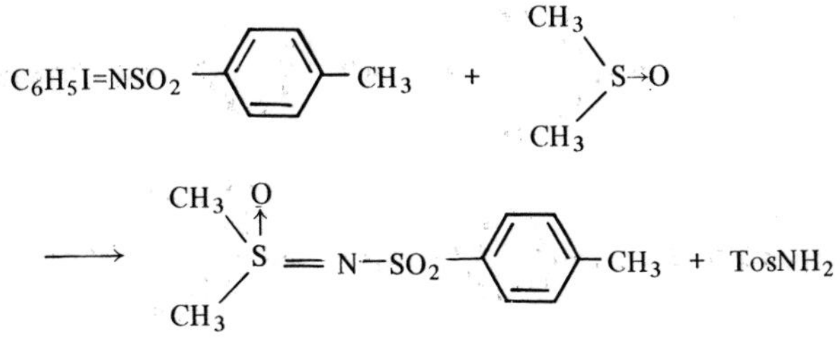

$$C_6H_5I{=}NSO_2-C_6H_4-CH_3 + (CH_3)_2S{\rightarrow}O$$

$$\longrightarrow (CH_3)_2S(\rightarrow O){=}N-SO_2-C_6H_4-CH_3 + TosNH_2$$

PART B

Cyclic Halonium Ions

CHAPTER 7

Ethylenehalonium Ions

7.1 PREPARATION BY NEIGHBORING HALOGEN PARTICIPATION AND PMR STUDY

Roberts and Kimbal[4] first suggested that the carbenium ion formed in the electrophilic attack of bromine on ethylene is stabilized by bridging by the neighboring β-bromine atom. They expected the ethylenebromonium ion to be slightly unsymmetric. Winstein and Lucas,[6] in subsequent studies on the stereochemical role of neighboring bromine in displacement reactions, concluded that the ethylenebromonium ion is either symmetric or that there is an extremely low barrier between the two unsymmetric forms. It is significant to point out that nothing in the original concept of neighboring-group participation, developed by Hughes and Ingold, and by Winstein,[153] requires momentary equivalence of the two carbon atoms involved in bromine bridging. It does require that an energy minimum or minima correspond to some degree of bonding of the carbocationic center to the neighboring halogen atom. Olah and Bollinger[9] reported in 1967 the first direct observation, by pmr spectroscopy, of tetramethylethylenehalonium ions under stable ion conditions in low nucleophilicity SbF_5-SO_2 solution. Subsequently, tri-, di-, monomethyl-, and even the parent ethylenehalonium ions were also obtained under related experimental conditions.[14] The pmr parameters of these ethylenehalonium ions are summarized in Table 34.

7.1.1 Parent Ethylenehalonium Ions

The parent ethylenhalonium ions (X = Br and I) were obtained when 1-halo-2-fluoroethanes were ionized in SbF_5-SO_2 solution at $-60°$.[14] The pmr spectra of the ethylenebromonium and iodonium ions show a singlet at δ 5.53 and

$$XCH_2CH_2F \xrightarrow{SbF_5\text{-}SO_2,\ -60°} \underset{\overset{+}{X}}{CH_2\text{—}CH_2} \quad SbF_6^-$$

X = Br, I

5.77, respectively. Under similar experimental conditions, when 1,2-dichloroethane and 1-chloro-2-fluoroethane were treated with SbF_5-SO_2 solution, donor-acceptor complexes were only formed instead of the ethylenechloronium ion.

$$SbF_5 \leftarrow ClCH_2CH_2Cl \qquad SbF_5 \leftarrow FCH_2CH_2Cl$$

Recently, however, the ethylenechloronium ion was also prepared.[154] When, instead of SO_2, SO_2ClF was used as the solvent, the reaction of antimony

TABLE 34

Pmr Data of Ethylenehalonium Ions[a]

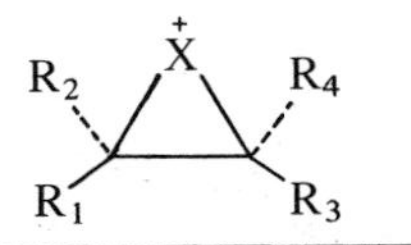

X	R_1	$\delta\ R_1$[b]	R_2	$\delta\ R_2$[b]	R_3	$\delta\ R_3$[b]	R_4	$\delta\ R_4$[b]
Cl	H	5.90 (s)	H	5.90 (s)	H	5.90 (s)	H	5.90 (s)
Br	H	5.53 (s)	H	5.53 (s)	H	5.53 (s)	H	5.53 (s)
I	H	5.77 (s)	H	5.77 (s)	H	5.77 (s)	H	5.77 (s)
Cl[c]	CH_3	3.00 (s)	H	6.24 (s)	H	6.24 (s)	H	6.24 (s)
Br	CH_3	2.98, (d) $J = 6$	H	7.84 (m)	H	5.86 (d)	H	5.86 (d)
I	CH_3	3.32, (d) $J = 6$	H	7.33 (m)	H	5.77 (d)	H	5.77 (d)
Br	CH_3	2.61 (m)	H	6.72 (m)	CH_3	2.61 (m)	H	6.72 (m)
Br	CH_3	2.61 (m)	H	6.72 (m)	H	6.72 (m)	CH_3	2.61 (m)
I	CH_3	3.07 (m)	H	6.87 (m)	H	6.87 (m)	CH_3	3.07 (m)
I	CH_3	3.07 (m)	H	7.14 (m)	CH_3	3.07 (m)	H	7.14 (m)
Br	CH_3	3.32 (s)	CH_3	3.52 (s)	H	5.55 (s)	H	5.55 (s)
I	CH_3	3.45 (s)	CH_3	3.45 (s)	H	5.72 (s)	H	5.72 (s)
Cl	CH_3	3.41 (s)	CH_3	3.41 (s)	CH_3	2.49, (d) $J = 6$	H	6.35, (q) $J = 6$
Br	CH_3	3.10 (s)	CH_3	3.10 (s)	CH_3	2.62, (d) $J = 6$	H	6.58, (q) $J = 6$

TABLE 34

Pmr Data of Ethylenehalonium Ions[a] (continued)

X	R_1	δR_1[b]	R_2	δR_2[b]	R_3	δR_3[b]	R_4	δR_4[b]
I	CH_3	3.23 (s)	CH_3	3.23 (s)	CH_3	2.92, (d) $J = 6.5$	H	6.75, (q) $J = 6.5$
Cl	CH_3	2.72 (s)	CH_3	2.72 (s)	CH_3	2.72 (s)	CH_3	2.72 (s)
Br	CH_3	2.86 (s)	CH_3	2.86 (s)	CH_3	2.86 (s)	CH_3	2.86 (s)
I	CH_3	3.05 (s)	CH_3	3.05 (s)	CH_3	3.05 (s)	CH_3	3.05 (s)

[a] Proton chemical shifts are referred to external capillary TMS in SbF_5-SO_2 solution at -60°.

[b] The multiplicities are shown in parentheses. s, singlet; d, doublet; q, quartet; m, multiplet.

[c] Rapidly equilibrating with open-chain ions, see text.

pentafluoride with 1-chloro-2-fluoroethane at $-80°$ gave a solution whose pmr spectrum consisted of these absorptions: a doublet (δ 4.6, 3H, J = 6 Hz), a quartet (δ 13.3 1H, J = 6 Hz), and a singlet (δ 5.9), consistent with the presence in the solution of the ethylenechloronium ions and methylchlorocarbenium ions, respectively.

$$ClCH_2CH_2F \xrightarrow[\text{1,2-H shift}]{SbF_5\text{-}SO_2ClF,\ -80°} \text{(ethylenechloronium ion)}\ SbF_6^- \;+\; CH_3\overset{+}{C}H\text{-}Cl\ SbF_6^- \leftrightarrow CH_3CH{=}\overset{+}{Cl}\ SbF_6^-$$

The ratio of these ions depends on the conditions under which they are prepared. If care is taken to keep the reactants below $-80°$, the ethylenechloronium ion predominates. Available data also show that the ions are not interconvertible. As the temperature is raised to $-50°$, the doublet and quartet of the ethylenechloronium ion disappear, and a new set of more shielded signals appears at δ 4.3 (doublet, 3H, J = 1.8 Hz) and δ 10.3 (quartet, 1H, J = 1.8 Hz), respectively, At the same time, the signal from the ethylenechloronium ion does not change its position and intensity. These data suggest the formation of the methylfluorocarbenium ion which is in equilibrium with 1,1-difluoroethane under the reaction conditions.[155]

$$CH_3\overset{+}{C}HCl \underset{SbF_5}{\overset{SbF_6^-}{\rightleftharpoons}} CH_3CHClF \underset{Cl^-}{\overset{SbF_5}{\rightleftharpoons}} CH_3\overset{+}{C}HF \underset{SbF_5}{\overset{SbF_6^-}{\rightleftharpoons}} CH_3CHF_2$$

7.1.2 Propylenehalonium Ions

When 2-fluoro-1-iodopropane and 2-fluoro-1-bromopropane were ionized in SbF_5-SO_2 solution at $-60°$, propyleneiodonium and propylenebromonium ions were formed, respectively.[14] The pmr spectrum of the propylenebromonium ion is shown in Figure 7. In an attempt to prepare trimethylenehalonium ions when 1,3-dihalopropanes were treated with SbF_5-SO_2 or SbF_5-FSO_3H-SO_2 solution at $-78°$ (see Section 0.0), the corresponding propylenehalonium ions were formed instead. The corresponding propylenechloronium ion could not

$$I(CH_2)_3X \xrightarrow{SbF_5\text{-}SO_2,\ -78°} \text{(propyleneiodonium ion: } CH_3(H)C\text{—}CH_2 \text{ bridged by } I^+)$$

X = Cl, I

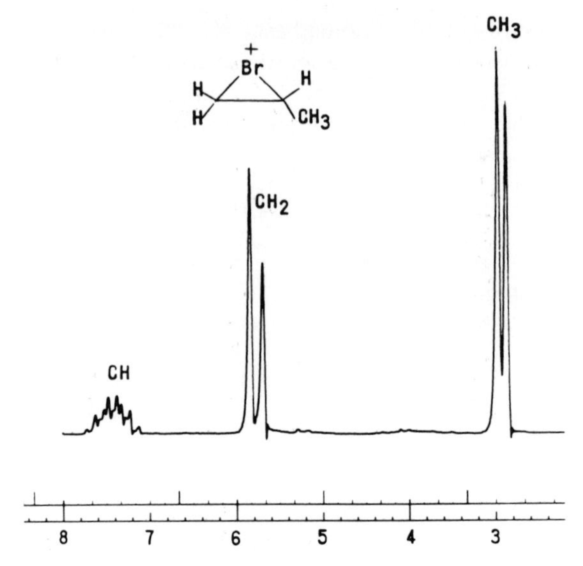

Figure 7. Pmr spectrum of the propylenebromonium ion.

$$Br(CH_2)_3X \xrightarrow[-78^\circ]{SbF_5\text{-}SO_2 \text{ or } SbF_5\text{-}FSO_3H\text{-}SO_2} \text{(propylenebromonium ion: } CH_3(H)C\text{—}C(H)H \text{ bridged by } Br^+)$$

$X = Cl, Br$

be obtained from dichloropropanes in SbF_5-SO_2 solution under similar conditions. Instead, when 1,1-, 1,2- or 1,3-dichloropropane was ionized in SbF_5-SO_2ClF solution at $-60°$; rapidly equilibrating ethylchlorocarbenium ions were formed. The equilibration processes is expected to also involve an unsymmetric bridged ion.

$$\begin{array}{l} Cl(CH_2)_3Cl \text{ or} \\ CH_3CHClCH_2Cl \text{ or} \\ CH_3CH_2CHCl_2 \end{array} \xrightarrow{SbF_5\text{-}SO_2ClF,\ -60^\circ} CH_3CH_2\overset{+}{C}HCl \rightleftharpoons CH_3\overset{+}{C}HCH_2Cl$$

$$\rightleftharpoons \text{(propylenechloronium ion: } CH_3(H)C\text{—}C(H)H \text{ bridged by } Cl^+)$$

It is of interest to compare ethylenechloronium and propylenechloronium ions. The former once formed are stable and are not transformed in the temperature range from −80 to −50° into methylchlorocarbenium ions. Thus the ethylenechloronium ion is a "static" ion, while the propylenechloronium ion is not.

7.1.3 2,3-Dimethylethylenehalonium Ions

Bridged 2,3-dimethylethylenehalonium ions were formed from *meso*- and *dl*-dibromobutanes, and from *erythro*- and *threo*-*dl*-2-bromo-3-fluorobutanes under stable ion conditions.[14] Starting from either conformational isomer in the brominated materials resulted in a mixture of two ions, that is, cis and trans halonium ions. The pmr spectra of the ions are shown in Figure 8.

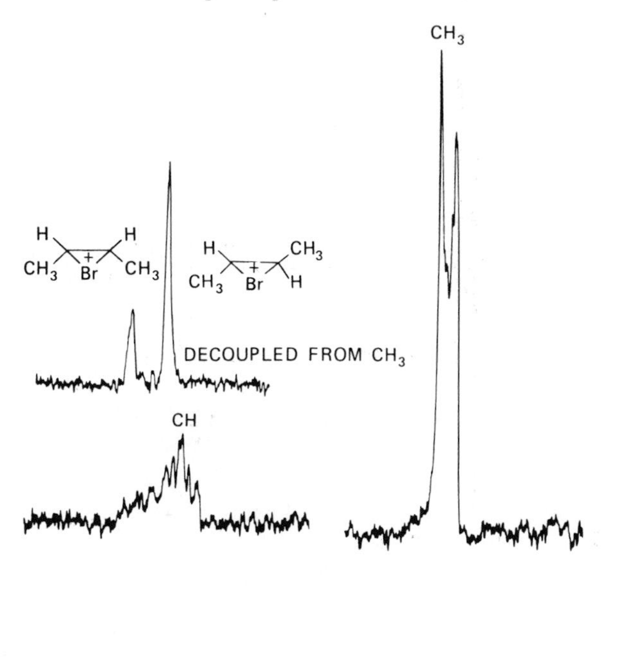

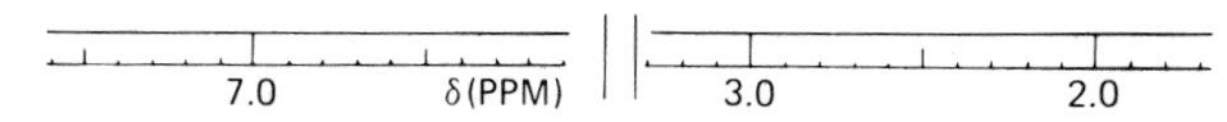

Figure 8. Pmr spectra of *cis*- and *trans*-2,3-dimethylethylenebromonium ions.

$CH_3CHBrCHFCH_3$ *erythro-dl-* or *threo-dl-*

$CH_3CHBrCHBrCH_3$ *meso-* or *dl-*

$\xrightarrow[-60 \text{ to } -80°]{SbF_5\text{-}SO_2}$

H, H, CH_3, CH_3, C—C, Br

CH_3, H, H, CH_3, C—C, $\overset{+}{Br}$

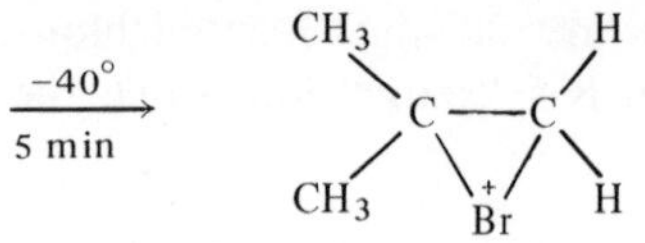

$$\xrightarrow[\text{5 min}]{-40^\circ} (CH_3)_2C\text{—}CH_2 \text{ (bridged by } Br^+)$$

The pmr spectra are complex (that of AA′XX′ system), but double irradiation decoupled them, revealing the presence of two (somewhat broadened) singlets in the methine region separated by 12 Hz. Regardless of the care used in the preparation of the ions, the same mixture was obtained. Furthermore, the same ratio of ions was also observed when *meso*- and *dl*-2,3-dibromobutane were treated with $CH_3F{\rightarrow}SbF_5\text{-}SO_2$ solution at −78°. Under these conditions

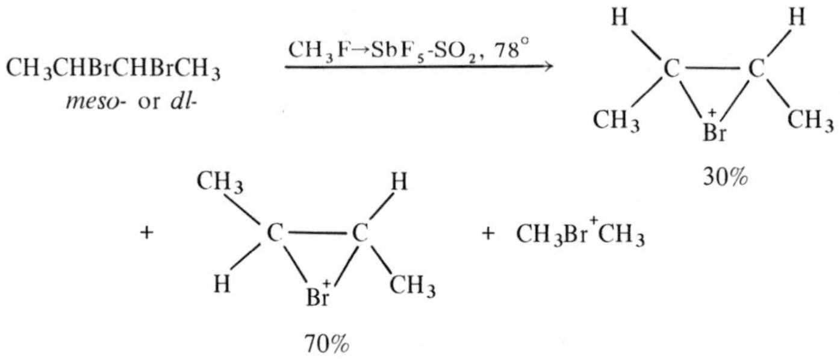

$$\underset{meso\text{- or }dl\text{-}}{CH_3CHBrCHBrCH_3} \xrightarrow{CH_3F\rightarrow SbF_5\text{-}SO_2,\ 78^\circ} \underset{30\%}{\text{cis-2,3-dimethylethylenebromonium ion}} + \underset{70\%}{\text{trans-2,3-dimethylethylenebromonium ion}} + CH_3Br^+CH_3$$

dimethylbromonium ion was also formed. Warming the isomeric 2,3-dimethylethylenebromonium ions to −40° for about 5 min causes isomerization to take place to give 2,2-dimethylethylenebromonium ions. The mechanism of this transformation is interesting. The breaking of a carbon-bromine bond must be followed by subsequent 1,2-hydrogen and 1,2-methyl shifts. The two involved intermediates were not observed, however, as long-lived ions.

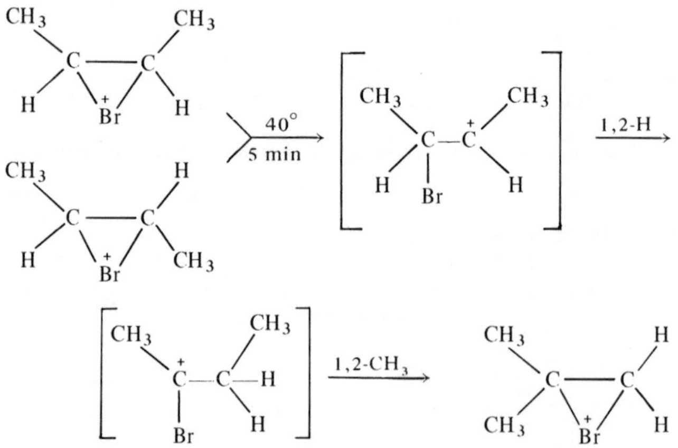

$$\text{cis- and trans-}CH_3CH\text{—}CHCH_3\ (Br^+) \xrightarrow[\text{5 min}]{40^\circ} [CH_3CHBr\text{—}\overset{+}{C}HCH_3] \xrightarrow{1,2\text{-H}} [CH_3\overset{+}{C}Br\text{—}CH_2CH_3] \xrightarrow{1,2\text{-}CH_3} (CH_3)_2C\text{—}CH_2\ (Br^+)$$

Ionization of *erythro-* or *threo-dl*-2-iodo-2-fluorobutane is at least 70% stereospecific. The ionization of *erythro-dl*-2-iodo-2-fluorobutane produces an ion to which trans configuration can only be assigned, but ionization of *threo-dl*-2-iodo-2-fluorobutane produces 70% cis-configuration ion and 30% *trans*-1,2-dimethylethylenehalonium ion. The structural assignments of the ions was based on the analysis of their pmr shifts and coupling constants (Figure 9),

$\xrightarrow{SbF_5SO_2,\ -60°}$ ~95% trans $\xrightarrow[K_2CO_3,\ -78°]{CH_3OH}$

erythro-dl-$CH_3CHICH(OCH_3)CH_3$

$\xrightarrow[-60°]{SbF_5\text{-}SO_2}$ 70% cis + 30% trans $\xrightarrow[K_2CO_3,\ -78°]{CH_3OH}$

threo-dl-$CH_3CHICH(OCH_3)CH_3$

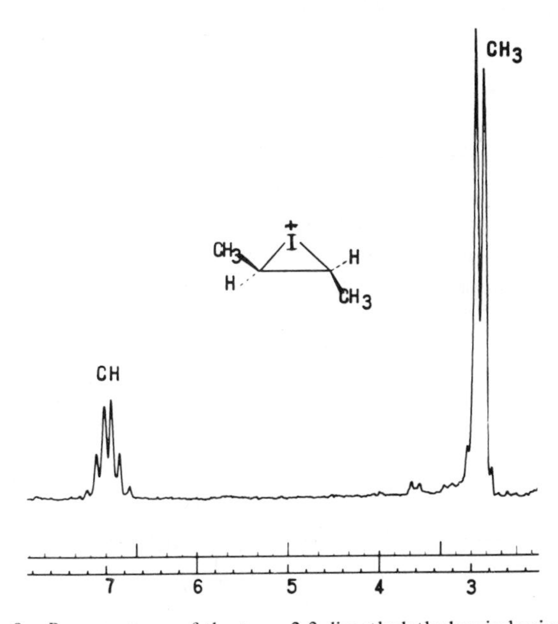

Figure 9. Pmr spectrum of the *trans*-2,3-dimethylethyleneiodonium ion.

as well as on methanolysis of the ions in the presence of solid potassium carbonate, which produced *erythro-dl*-2-iodo-3-methoxybutane and *threo-dl*-2-iodo-3-methoxybutane, respectively. Surprisingly, no corresponding rearrangement of the 2,3-dimethylethyleneiodonium ion to the 2,2-dimethylethyleneiodonium ion, or interconversion of the stereosiomers, could be detected after 10 min at -15°, although prolonged storage at about this temperature results in some conversion of each isomer to other.

Ionization of *erythro-* or *threo*-2-fluoro-3-chlorobutane or *meso* or *dl*-2,3-dichlorobutane results in immediate rearrangement to a mixture of dimethyl-, chloromethyl-, and methy, ethylchlorocarbenium ions in a 40:60 ratio.

$$CH_3CHClCHClCH_3 \xrightarrow{SbF_5\text{-}SO_2,\ -60^\circ} \quad \text{or} \quad CH_3CHClCHFCH_3 \xrightarrow{SbF_5\text{-}SO_2,\ -60^\circ} \quad (CH_3)_2\overset{+}{C}\text{—}CH_2Cl\ (40\%) + CH_3CH_2\overset{+}{C}(CH_3)Cl\ (60\%)$$

7.1.4 2,2-Dimethylethylenehalonium Ions

When 1-halo-2-fluoromethylpropanes are ionized in SbF_5SO_2 solution at -78 °, dimethylethylenehalonium ions are formed with bromine and iodine

$$CH_3\text{–}C(CH_3)(F)\text{–}CH_2X \xrightarrow{SbF_5\text{-}SO_2,\ -78^\circ} (CH_3)_2C\text{—}CH_2\ (\text{bridged by } \overset{+}{X})\quad SbF_6^-$$

X = Br, I X = Br, I

acting as *n*-donor participating atoms.[14c] The pmr spectra for the bromo- and iodonium ions are similar, and each displays two singlet absorptions at δ 3.32 (3H) and 5.55 (1H) or δ 3.45 (3H) and 5.72 (1H), respectively. Chlorine and fluorine, however, show no ability to form bridged ions, and the corresponding ions exist solely as open-chain halocarbenium ions.

$$CH_3\text{–}C(CH_3)(F)\text{–}CH_2Cl \xrightarrow{SbF_5\text{-}SO_2,\ -78^\circ} (CH_3)_2\overset{+}{C}\text{–}CH_2Cl \xrightarrow[-78^\circ]{\text{Slow}} CH_3CH_2\overset{+}{C}(CH_3)Cl$$

$$CH_3\text{–}C(CH_3)(OH)\text{–}CH_2F \xrightarrow{SbF_5\text{-}FSO_3H\text{-}SO_2,\ -80^\circ} CH_3\text{–}C(CH_3)(\overset{+}{O}H_2)\text{–}CH_2F \xrightarrow[-80^\circ]{-H_2O} CH_3CH_2\overset{+}{C}(CH_3)F$$

Furthermore, protonation of 1-iodo-2-methoxy-2-methylpropane in SbF_5-FSO_3H-SO_2 solution immediately gives the 2,2-dimethylethyleneiodonium ion and protonated methanol, while the corresponding chloride and bromide are

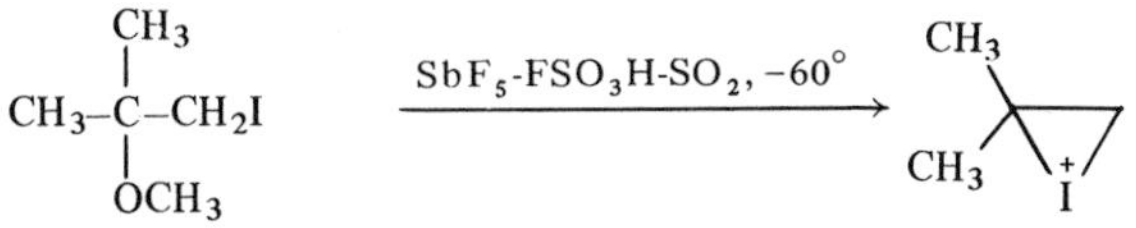

$$CH_3-C(CH_3)(OCH_3)-CH_2I \xrightarrow{SbF_5\text{-}FSO_3H\text{-}SO_2,\ -60^\circ} (CH_3)_2C\overset{+}{I}CH_2$$

O-protonated and then slowly decompose to unidentified products (polymers).

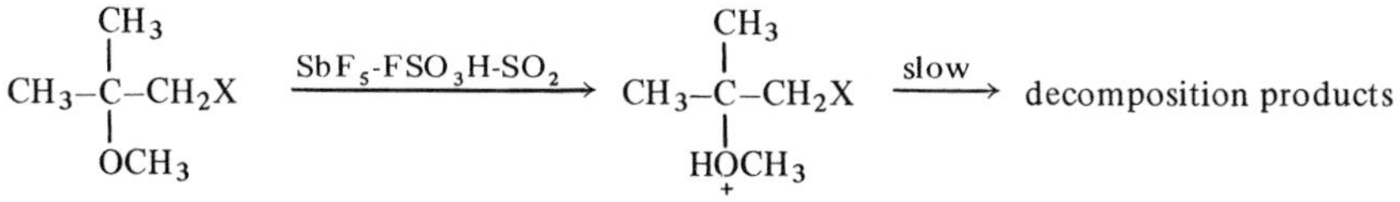

$$CH_3-C(CH_3)(OCH_3)-CH_2X \xrightarrow{SbF_5\text{-}FSO_3H\text{-}SO_2} CH_3-C(CH_3)(H\overset{+}{O}CH_3)-CH_2X \xrightarrow{\text{slow}} \text{decomposition products}$$

X = Cl, Br

7.1.5 Trimethylethylenehalonium Ions

When 2-fluoro-3-chloro(or bromo or iodo)-2-methylbutane is ionized in SbF_5-SO_2 solution at -78°, stable trimethylethylenehalonium ions (X = Cl, Br, I) are formed.[14a] The pmr data of the ions are summarized in Table 34. The

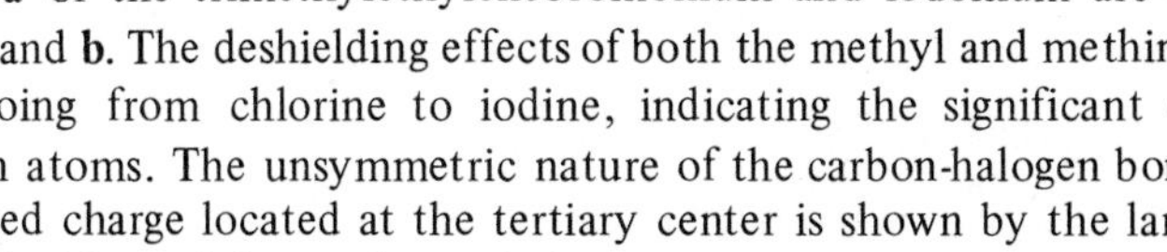

$$CH_3-C(CH_3)(F)-CH(X)CH_3 \xrightarrow{SbF_5\text{-}SO_2,\ -78^\circ} (CH_3)_2C\overset{+}{X}CH(CH_3)$$

X = Cl, Br, I

pmr spectra of the trimethylethylenebromonium and iodonium are shown in Figure 10**a** and **b**. The deshielding effects of both the methyl and methine protons decrease going from chlorine to iodine, indicating the significant charge at the halogen atoms. The unsymmetric nature of the carbon-halogen bonding and the increased charge located at the tertiary center is shown by the large difference in deshielding of the single methyl group compared to the geminal methyl groups.

In the spectra of trimethylethylenebromonium and -chloronium ions, equivalence of the geminal methyl groups is observed. This suggests either a rapid equilibration involving the open chain halocarbenium ions

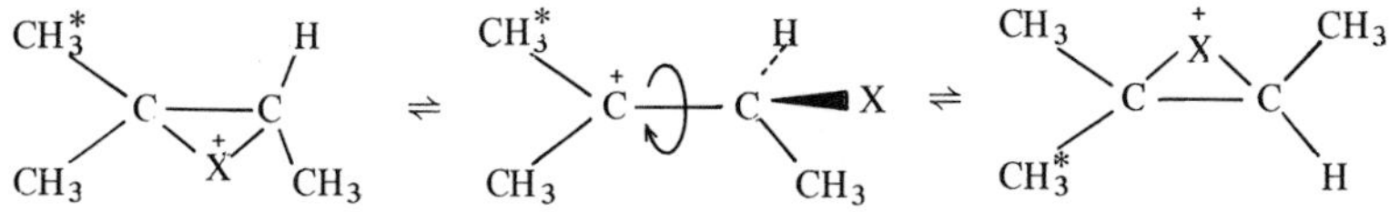

$$(CH_3^*)(CH_3)C\overset{+}{X}C(H)(CH_3) \rightleftharpoons (CH_3^*)(CH_3)\overset{+}{C}-C(H)(X)(CH_3) \rightleftharpoons (CH_3)(CH_3^*)C\overset{+}{X}C(CH_3)(H)$$

X = Cl, Br

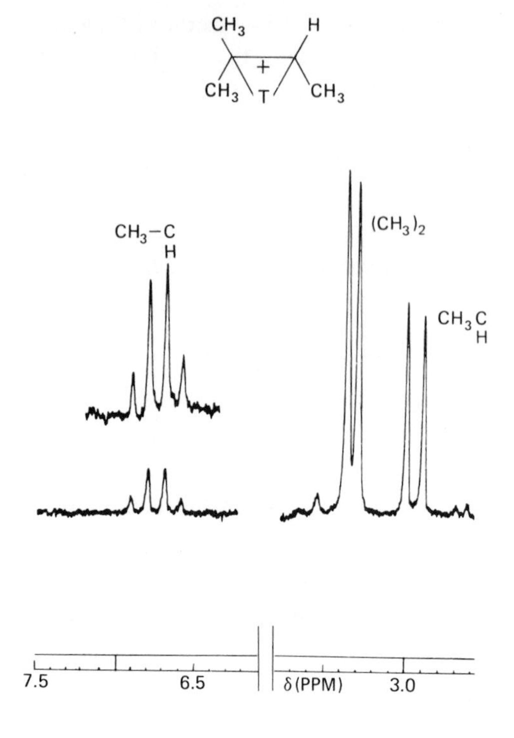

Figure 10a. Pmr spectrum of the trimethylethyleneiodonium ion.

or that the chemical shifts of the geminal methyl groups are accidentally equivalent. The former suggestion could be substantiated by the variation of the spectra of the ions with temperature, but in the temperature range so far studied this was not observed. The more suitable cmr spectra were not studied, however yet in sufficient detail.

7.1.6 Tetramethylethylenehalonium Ions

The first direct observation of ethylenehalonium ions in solution was made by Olah and Bollinger[9] in 1967, when 2,3-dihalo-2,3-dimethylbutanes were ionized in SbF_5-SO_2 solution at $-60°$. Stable tetramethylethylenehalonium ions were formed with chlorine, bromine, and iodine acting as donor atoms.

$$(CH_3)_2CX\text{—}CY(CH_3)_2 \xrightarrow{SbF_5\text{-}SO_2,\ -60°} (CH_3)_2C\text{—}C(CH_3)_2 \text{ (bridged by } \overset{+}{X}\text{)}$$

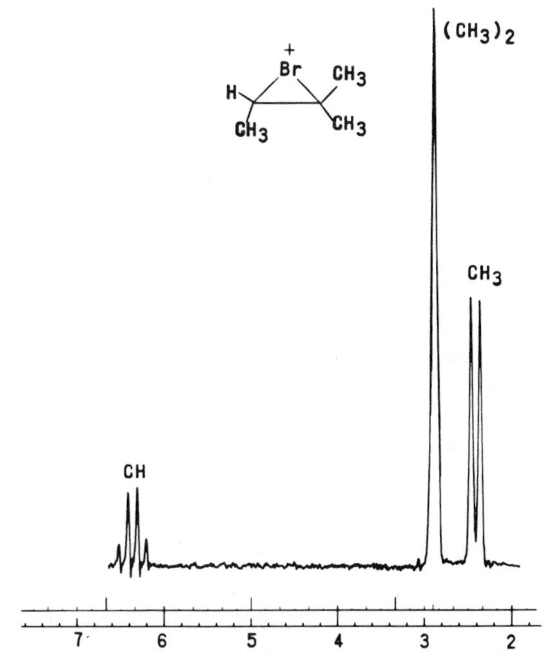

Figure 10b. Pmr spectrum of the trimethylethylenebromonium ion.

X = Cl; Y = F, Cl	X = Cl
X = Br; Y = F, Br	X = Br
X = I; Y = F	X = I

Fluorine as a neighboring group, however, gave only the rapidly equilibrating open-chain β-fluorocarbenium ion, even at $-90°$.

$$(CH_3)_2CF{-}CF(CH_3)_2 \xrightarrow{SbF_5\text{-}SO_2,\ -90°} (CH_3)_2CF{-}\overset{+}{C}(CH_3)_2 \rightleftharpoons (CH_3)_2\overset{+}{C}{-}CF(CH_3)_2$$

Whether this equilibration process involves a higher-lying, fluorine-bridged transition state (or intermediate) not yet observed, as suggested by Clark,[156]

$$(CH_3)_2C{-}C(CH_3)_2 \text{ bridged by } \overset{+}{F}$$

or, more probably, as suggested by Olah and Bollinger,[9] an equilibration process including 1,2-methyl shifts (in their ability to stabilize tertiary carbonium centers methyl and fluorine are not too different), cannot be determined by

$$(CH_3)(F)\overset{}{C}(CH_3)-\overset{+}{C}(CH_3)_2 \rightleftharpoons (CH_3)(F)\overset{+}{C}-C(CH_3)_3 \rightleftharpoons (CH_3)(F)C(CH_3)-\overset{+}{C}(CH_3)_2, \text{ etc.}$$

direct experimental evidence. However, in solution chemistry there seems to be no indication of the formation of any positively charged fluorine species. Bridged fluoronium ions therefore also seem to be unlikely, and the latter equilibrium seems to well explain the experimental data without the necessity of involving a bridged fluoronium ion.

The pmr spectra of tetramethylethylenehalonium ions show singlet absorptions at δ 2.72 (X = Cl), 2.86 (X = Br), and 3.05 (X = I). The rapidly equibrating fluorinated ion, in contrast, shows a doublet at δ 3.10 (J_{HF} = 11 Hz) in its pmr spectrum.

2,2-Dichloro-3,3-dimethylbutane (pinacolone dichloride) is also ionized in SbF_5-SO_2 solution at $-60°$ to give the tetramethylethylenechloronium ion formed through the unstable chlorocarbenium ion intermediate. Ionization

$$(CH_3)_3C-CCl_2-CH_3 \xrightarrow{SbF_5\text{-}SO_2,\ -60°} (CH_3)_3C-\overset{+}{C}(Cl)CH_3 \longrightarrow (CH_3)_2C\overset{\overset{+}{Cl}}{—}C(CH_3)_2$$

of 2-halo-2-methoxy- and 2-halo-3-acetoxy-2,3-dimethylbutanes in either SbF_5-SO_2 or SbF_5-FSO_3H-SO_2 solution yields a mixture of tetramethylethylenehalonium ions, acetoxonium ions, and methylated ketones.

$$(CH_3)_2C(X)-C(CH_3)_2(OCH_3) \xrightarrow[\text{or } SbF_5\text{-}SO_2]{FSO_3H\text{-}SbF_5\text{-}SO_2} (CH_3)_2C\overset{\overset{+}{X}}{—}C(CH_3)_2 + CH_3\overset{+}{O}H_2 \quad \text{and} \quad (CH_3)_3C-C(CH_3)=\overset{+}{O}CH_3$$

X = Cl, Br, I

$$(CH_3)_2C(X)\text{—}C(CH_3)_2OOCCH_3 \xrightarrow[\text{or } SbF_5\text{-}SO_2]{FSO_3H\text{-}SbF_5\text{-}SO_2} (CH_3)_2C\text{—}C(CH_3)_2 \text{ (bridged } \overset{+}{X}\text{)} + CH_3CO_2H_2^+$$

X = Cl, Br, I

7.1.7 1-Butyleneiodonium Ion (2-Ethylethyleneiodonium Ion)

Olah and Peterson reported the preparation of the 1-butyleneiodonium ion from the ionization of either 1-fluoro-4-iodobutane or 1-iodo-2-fluorobutane

$$FCH_2(CH_2)_2CH_2I \xrightarrow{SbF_5\text{-}SO_2}$$

$$ICH_2CHFCH_2CH_3 \xrightarrow{SbF_5\text{-}SO_2,\ -78^\circ} \overset{+}{I}\text{-bridged}\text{—}CH_2CH_3$$

with SbF_5-SO_2 solution at $-78°$.[15] Tetramethyleneiodonium ion was also formed in the reaction of 1-fluoro-4-iodobutane. The mechanism of its formation may be the following.

$$ICH_2CH_2CH_2CH_2F \xrightarrow{SbF_5\text{-}SO_2} (ICH_2CH_2CH_2\overset{+}{C}H_2) \rightarrow ICH_2CH_2C^+HCH_3$$

$$ICH_2\overset{+}{C}(CH_3)CH_3 \nleftarrow ICH_2C^+HCH_2CH_3 \rightarrow \overset{+}{I}\text{-bridged}\text{—}C_2H_5$$

Furthermore, ionization of 1-iodo-3-halobutanes in SbF_5-SO_2 solution at $-60°$ gives only the 1-butyleneiodonium ion. Obviously, ionization takes place at the secondary carbon atom to give an intermediate ion which then undergoes a 1,2-hydrogen shift. Subsequent iodine participation would form the stable 1-butyleneiodonium ion. However, when 1,2- and 1,3-dibromo(or chloro)butanes

$$\underset{\substack{| \quad\quad\ \ | \\ X \quad\quad\ \ I}}{CH_3CHCH_2CH_2} \xrightarrow[-60^\circ]{SbF_5\text{-}SO_2} \underset{\substack{| \\ I}}{CH_3\overset{+}{C}HCH_2CH_2} \xrightarrow{1,2\text{-}H^-} \underset{\substack{| \\ I}}{CH_3CH_2\overset{+}{C}HCH_2} \longrightarrow$$

$X = Cl, Br, I$

are treated with SbF_5-SO_2 solution under similar conditions, only the corresponding tetramethylenehalonium ions are obtained. Their formation is an unusual reaction for a cationic species in strong acid media in which one

$$\left.\begin{array}{l} CH_3CHXCH_2CH_2X \\ \\ CH_3CH_2CHXCH_2X \end{array}\right. \xrightarrow{SbF_5\text{-}SO_2,\ -70^\circ} \text{(five-membered ring, } \overset{+}{X}\text{)} \qquad X = Cl, Br$$

usually finds nonbranched cations rearranging to branched ones, but rarely the reverse. The driving force for this rearrangement is assumed to be the formation of a very stable five-membered-ring ion. The mechanism for this rearrangement

$$CH_3CH_2\overset{+}{C}HCH_2Br \longrightarrow CH_3\overset{+}{C}HCH_2CH_2Br \longrightarrow \overset{+}{C}H_2CH_2CH_2CH_2Br \longrightarrow \text{(five-membered ring, } \overset{+}{Br}\text{)}$$

may involve successive 1,2-hydrogen shifts. Thus far, 1-butylenebromonium or -chloronium ions are not yet known.

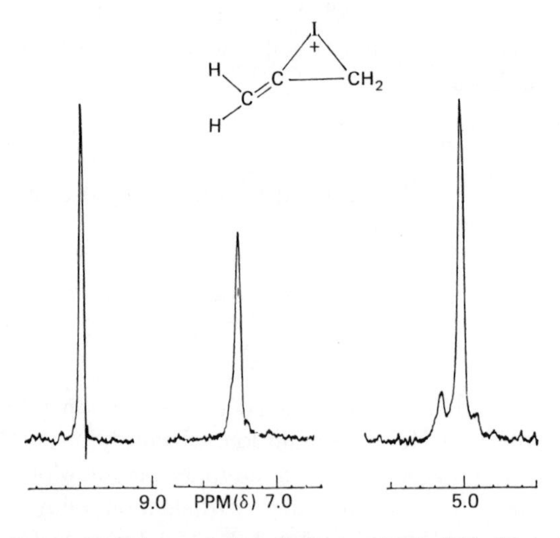

Figure 11. Pmr spectrum of the propadienyliodonium ion in SbF_5-SO_2 solution at -80°.

7.1.8 Propadienylhalonium Ions

Propadienylhalonium ions were prepared from 2,3-dihalopropenes and SbF_5-SO_2 solution at $-70°$.[157] The pmr spectrum of the propodienyliodonium ion is

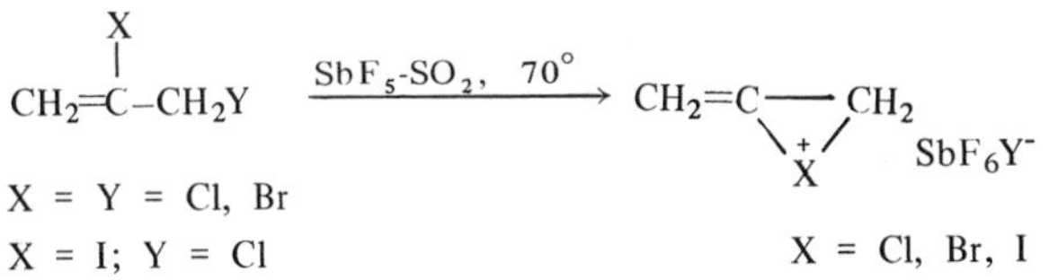

$$CH_2{=}C(X){-}CH_2Y \xrightarrow{SbF_5\text{-}SO_2,\ 70°} CH_2{=}C\underset{\overset{+}{X}}{\diagdown\!\diagup}CH_2 \quad SbF_6Y^-$$

X = Y = Cl, Br

X = I; Y = Cl

X = Cl, Br, I

shown in Figure 11. Pmr data of the ions in each case show three singlets with a ratio of 2:1:1 (from high field to low field). Geminal and long-range couplings are not observed in the ions.

7.2 CARBON-13 NMR STUDIES OF ETHYLENEHALONIUM IONS

Olah and White[158] first reported carbon-13 nmr studies of ethylenehalonium ions including that of the tetramethylethylenebromonium ion. A recent more detailed reinvestigation of the cmr of ethylenehalonium ions included

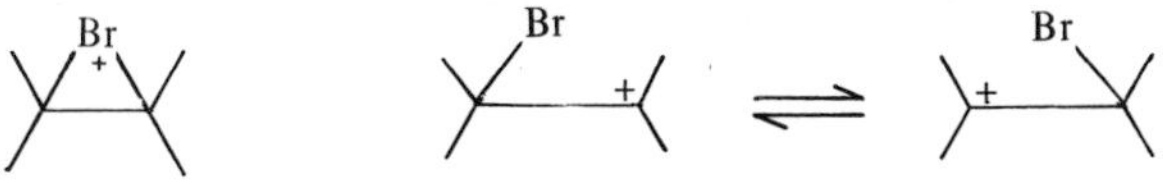

Br Br Br

Symmetrically bridged ion

Equilibrating open-chain halocarbenium ions

a comprehensive series of ethylenebromonium, as well as ethylenechloronium and -iodonium ions.[26d] The cmr data of the ethylenehalonium ions are summarized in Table 35.

The structures of most of the ions in Table 35 have been discussed previously in connection with their pmr spectral parameters. Cmr studies, however, give valuable further information.

There are several structural possibilities for ethylenehalonium ions: the static symmetrically bridged ion in which the C_2–X bond length is equal to the C_3–X bond length (**1**), the static unsymmetrically bridged ion in which the C_2–X bond length is not equal to the C_3–X bond length (**2**), and the halonium ion in equilibrium with a pair of open-chain β-haloalkenium ions (**3**). If one

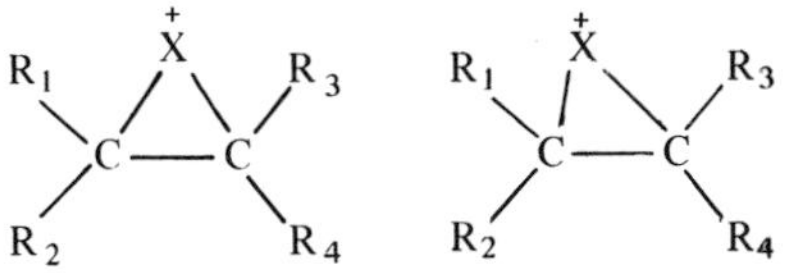

$\overset{+}{X}$ R_1 R_3 C—C R_2 R_4 $\overset{+}{X}$ R_1 R_3 C—C R_2 R_4

1 **2**

TABLE 35
Carbon-13 Chemical Shifts of Ethylenehalonium Ions[a]

R_1	R_2	R_3	R_4	X	C-2	C-3	CH_3 (C-2)	CH_3 (C-3)
H	H	H	H	Br	73	73		
CH_3	H	H	H	Br	122.2	72.7	25.4	
CH_3	H	CH_3	H	Br[b]	110.9	111.8	22.4	22.4
H	CH_3	CH_3	H	Br	108.8	108.8	17.4	17.4
CH_3	CH_3	H	H	Br	211.4	50.5	35.4	
CH_3	CH_3	CH_3	H	Br	172.8	92.7	30.5	18.2
CH_3	CH_3	CH_3	CH_3	Br	139.7	139.7	29.7	26.7
H	H	H	H	Cl[c]	74.1	74.1		
CH_3	CH_3	CH_3	CH_3	Cl	151.7	151.7	28.0	28.0
H	H	H	H	I	87.9	67.9		
CH_3	H	CH_3	H	I	105.2	105.2	24.3	24.3
H	CH_3	CH_3	H	I	104.0	104.0	19.2	19.2
CH_3	CH_3	H	H	I	186.0	52.8	33.2	
CH_3	CH_3	CH_3	CH_3	I	124.8	124.8	28.0	28.0

[a] parts per million from CS_2. In SO_2 at −40° unless otherwise indicated.
[b] In SO_2 at −60°.
[c] In SO_2ClF at −80°.

$$\begin{matrix} R_1 \\ \diagdown \\ \end{matrix} \overset{+}{C} - C(X)(R_4) - R_3 \rightleftharpoons \underset{\mathbf{3}}{R_1R_2C \overset{\overset{+}{X}}{\frown} CR_3R_4} \rightleftharpoons R_1 - C(X)(R_2) - \overset{+}{C}R_3R_4$$

of the open ions is a primary carbenium ion its formation will be energetically particularly unfavorable.

7.2.1 Symmetrically Substituted Ethylenehalonium Ions

For symmetrically substituted ions only structures **1** and **3** are possible, and in the latter case there would be equal amounts of the two open-ion forms.

Both chemical shift and carbon-hydrogen coupling constant data show that the structure of the ethylenebromonium ion is that of the bridged ion **1**. Similar considerations (i.e., chemical shift data) applied to the other parent ethylene-halonium ions indicate that they also are symmetrically bridged ions. The carbon-13 chemical shifts of the three parent ethylenehalonium ions are in the range $\delta_{^{13}C}$ 74-68, whereas the α-carbon shifts of three related open-chain di-alkylhalonium ions occur over a much wider range. This most likely arises from a larger β-substituent effect in ethylene halonium ions, because of the greater proximity of the positively charged halogen atom to the β carbon in these cyclic ions (C_β–C_α–X). The carbon shift is dominated by charge effects. rather than by the individual differences in the nature of the halogen atoms. The large β-substituent effect in the cyclic ions explains why attempts to estimate their chemical shifts by regarding each carbon atom as subject to both an α- and β-substituent effect are unsuccessful (except in the case of the ethylene-chloronium ion, in which the agreement may be fortuitous).

The cmr and pmr spectra of the 2,3-dimethylethylenebromonium ion, pre-pared from either *meso* or *dl*-dibromobutanes, reveal the presence of cis and trans isomers. The ion is therefore most likely the symmetrically bridged ion, since it would be extraordinary if noticeable amounts of open ion did not lead to the interconversion of cis and trans isomers at a rate considered rapid on the nmr time scale.

The 5-ppm upfield shift in the methyl carbon shielding of the cis isomer compared with the trans isomer reflects the larger steric interactions between the methyl groups in the former case.[159] A similar effect of the same magnitude is observed in the two isomers of the 1,2-dimethylethyleneiodonium ion.

The cis/trans ratio in the 2,3-dimethylethyleneiodonium ion depends on which precursor is used to prepare the ion. The iodonium ion, then, clearly is a symmetrically bridged ion.

The structure of the tetramethylethylenebromonium ion has been discussed in some detail.[3,14,26,158] The cmr data first were considered inconsistent with either a bridged species or a rapidly equilibrating pair of open ions, so the structure of the ion was proposed to be a mixture of both forms.

However, the available nmr data are still subject to two interpretations. The larger than usual substituent effect[153,160,161] that occurs in the C-2 and C-3 shielding of the ethylenebromonium ion on methyl substitution may indicate an increased carbenium ion character for the C-2 and C-3 carbons if the ion is a static bridged ion, or some contribution of open-chain β-halocarbenium ions if the ion is a symmetrically bridged ion in equilibrium with open-ions forms. In support of the former structure, it has been argued from carbon-13 chemical shift data for the tetramethylethylenephenonium ion that the ion has the symmetrically bridged structure.

It is of interest that the differences in the ring carbon shifts of the three tetramethylethylenehalonium ions are larger than in the three parent ions. Iodine can accommodate positive charge better than chlorine or bromine, so more carbenium ion character at C-2 and C-3 exists in the tetramethylethylenebronomium or -chloronium ion than in -iodonium ion, or there is less contribution of the β-halocarbenium ion for iodine than for chlorine or bromine.

7.2.2 Unsymmetrically Substituted Ethylenehalonium Ions

2-Methyl- and 2,2-dimethylethylenebromonium ions may exist as unsymmetrically bridged ions (2) or as equilibrium mixtures of unsymmetrically bridged ions and open-chain secondary or tertiary haloalkyl cations. On the basis of cmr data for one particular temperature alone it is not possible to decide between these two possibilities.[162]

The α-substituent effect of the methyl group in the 1-methyl-1,1-dimethylethylenehalonium ion is 49 ppm, while that of two geminal methyl groups in the 1,1-dimethyl- ion is 138.4 ppm. Accompanying these downfield shifts at C-1 are upfield shifts of 0.3 and 12.5 ppm for the C-2 shielding in the two ions, respectively.

The data for these ions, and those for β-bromocumyl ions, show β-bromine substituent effects, in ions of the type $RR'C^+CH_2Br$, of from 21 to 117 ppm. Apparently, the extent to which bromine may participate in the stabilization of a positively charged β carbon may vary, depending on the electron needs of the carbon. The cmr spectrum of the 2,2-dimethylethylenebronomium ion was studied at different temperatures to try to distinguish between a static unsymmetrically bridged ion whose cmr spectrum should be temperature-independent and an equilibrating system of a bridged ion with a β-halocarbenium ion whose cmr spectrum should vary with temperature. The 2,2-dimethyl-

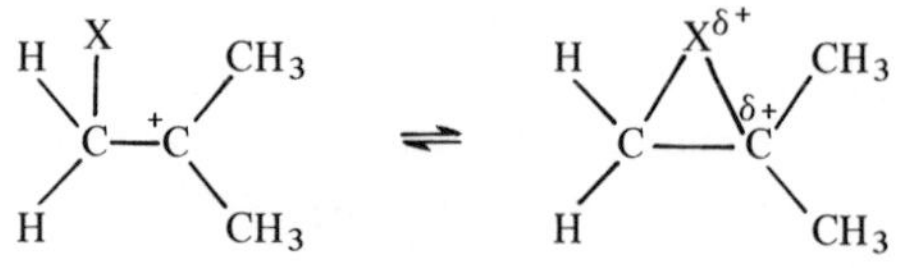

$$H_2C(X)\text{—}C^+(CH_3)_2 \rightleftharpoons H_2C\text{—}C^{\delta+}(CH_3)_2 \text{ (bridged by } X^{\delta+})$$

ethylenechloronium ion has been shown to exist as a β-haloalkylcarbenium ion. Since bromine has a greater ability than chlorine to accommodate a positive charge, it is excepted that the corresponding bromonium ion will be intermediate in character between a purely bridged ion and a β-haloalkylcarbenium ion, and therefore its cmr spectrum should be quite dependent on temperature.

The carbon-13 chemical shifts for the methylene and methyl carbons in the ion at several temperatures (−20 to −80°) are shown in Table 36. It

TABLE 36
Temperature Dependence of the Carbon-13 Chemical Shifts in the 2,2-Dimethylethylenebromonium Ion

Temperature (°C)[b]	$\delta_{^{13}C}$[a]			
	C-2	CH_2	CH_3	Δ
−20		59.10	34.21	24.89
−30		59.27	34.19	25.08
−40	211.3	59.37	33.98	25.38
−50		59.62	34.05	25.56
−60		59.58	33.71	25.87
−70		60.09	33.76	26.34
−80	206.17	59.87	33.45	26.41

[a] In SO_2; parts per million from external capillary of TMS.
[b] ±1°.

was impractical to obtain the C-2 shieldings for all seven temperatures in Table 36, because of the vary large number of FT accumulations necessary to observe this signal at the low ion concentrations obtainable for this low-temperature study. Signals were measured from external TMS, and it is not known how significant the small absolute changes for δ_{CH_3} and δ_{CH_3} shown in Table 36 are. However, the internal chemical shift difference ($\delta_{CCH_3} - \delta_{CCH_2}$) should be temperature-independent, so that changes in its value reflect changes in the position of the equilibrium. An increase in the open-ion form will result in a deshielding of the CH_3 and C^+ resonances, but it is not obvious what should happen to the CH_2 absorption. In the open-ion form the methylene carbon shielding is subject to α-bromine and α-C^+ substituent effects, while in the bridged ion it experiences a partial α-C^+ effect and an increased bromine substituent effect, since the bromine now carries more charge. (For the analogous ion in the five-membered-ring series this problem does not exist, because in the open ion the shielding of the methylene carbon bonded to bromine is not subject to an α-C^+ effect, hence should be shielded from the corresponding resonance in a bridged structure.) The observed variation in $\delta_{CH_3} - \delta_{CH_2}$ with temperature therfore cannot be given a detailed explanation in terms of of molecular processes, but it does signify some sort of equilibrium.

The deshielding of 5 ppm in the C-2 shielding with a 40° decrease in temperature indicates that at lower temperatures more of the bridged ion is present than at higher temperatures. The relatively small change in the C-2 resonance suggests that the 2,2-dimethylethylenebromonium ion is mainly the static bridged ion in equilibrium with only a small amount of the β-haloalkylcarbenium ion.

Substitution of a third methyl group into the ethylenebromonium ion results in an ion whose charge distribution is more symmetrically disposed. This is reflected by an upfield shift of 39 ppm in the C-2 shielding compared with dimethylethylenebromonium ions. The equivalency of the geminal methyl groups in both the pmr and the cmr spectra of the 2,2,3-trimethylethylenehalonium ion suggests that the ion exists as an equilibrating mixture of the unsymmetrically bridged ion and an open-ion form. It is, however, possible, although less likely, because of the large cis methyl shielding effect, that both the geminal methyl carbon and proton shielding are accidentally equivalent in the bridged ion, and that consequently equilibration does not occur.

$(CH_3)_2\overset{3}{C}{-}\overset{2}{C}H_2$ bridged by $\overset{+}{Br}$; CH_3 35.8

$(CH_3)_2\overset{3}{C}{-}\overset{2}{C}(CH_3)H$ bridged by $\overset{+}{Br}$; CH_3 30.8

$(CH_3)_2\overset{3}{C}{-}\overset{2}{C}(CH_3)_2$ bridged by $\overset{+}{Br}$; CH_3 26.8

$CH_3(H)\overset{3}{C}{-}\overset{2}{C}H_2$ bridged by $\overset{+}{Br}$; CH_3 25.8

$CH_3(H)\overset{3}{C}{-}\overset{2}{C}(H)CH_3$ bridged by $\overset{+}{Br}$; CH_3 22.8

$CH_3(H)\overset{3}{C}{-}\overset{2}{C}(CH_3)_2$ bridged by $\overset{+}{Br}$; CH_3 17.8

The order of the methyl shieldings is the same as we would predict from a consideration of the relative degrees of electron deficiency at C-2 and C-3 in these ions; similarly for the methyl shifts in related ions grows. For corresponding ions in both series the methyl shift is deshielded in the former case, as expected.

7.3 PREPARATION VIA PROTONATION OF CYCLOPROPYL HALIDES

The protonation of cyclopropyl halides in superacids gives the corresponding three-membered-ring halonium ions.[163] The highly substituted ethylenehalonium ions, on warming to −10°, dehydrohalogenate to give the corresponding substituted allyl cations.

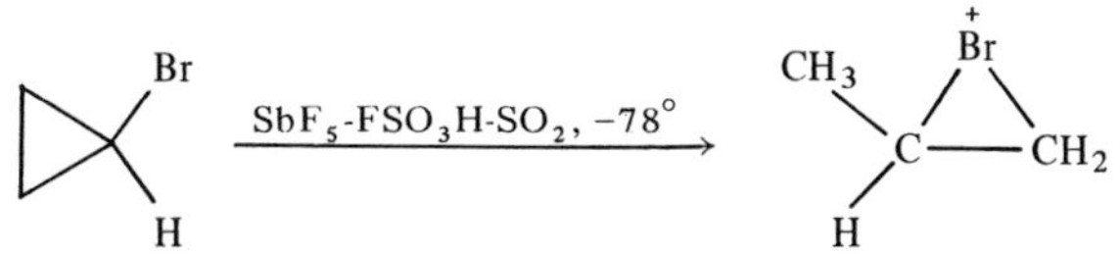

SbF_5-FSO_3H-SO_2, −78°

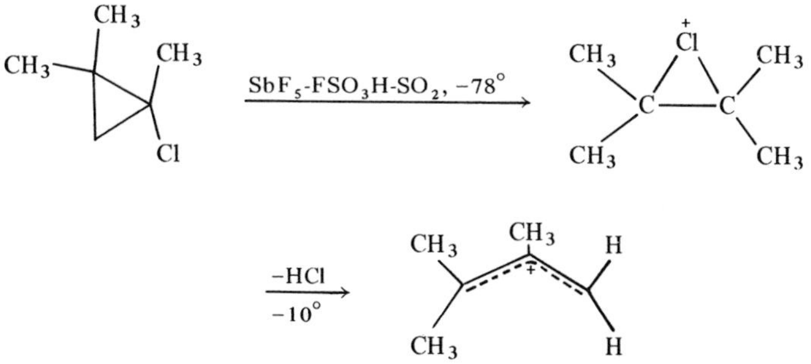

SbF_5-FSO_3H-SO_2, −78°

−HCl, −10°

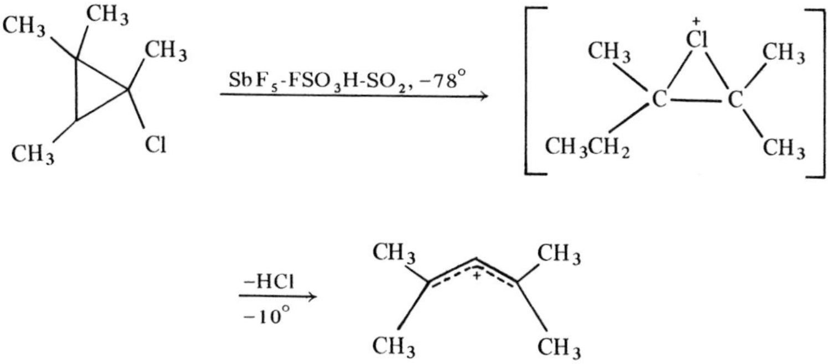

SbF_5-FSO_3H-SO_2, −78°

−HCl, −10°

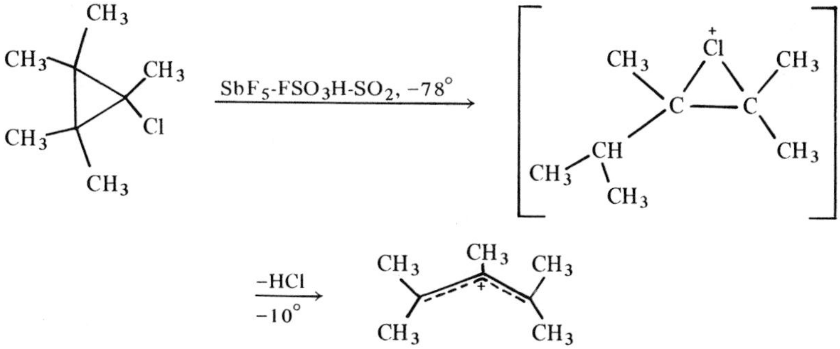

SbF_5-FSO_3H-SO_2, −78°

−HCl, −10°

7.4 PREPARATION VIA DIRECT HALOGENATION OF OLEFINS

We have already discussed the preparation of ethylenehalonium ions from 1,2-dihaloalkanes and SbF_5-SO_2 under stable ion conditions. Electrophilic additions of halogens to olefins show high trans stereoselectivity, and consequently are assumed to take place via ethylenehalonium ion intermediates.[4-7,164] When elementary halogens are reacted with olefins in the presence of Lewis acids (e.g., SbF_5), ethylenehalonium ions are formed. Even though the corresponding ethylenehalonium ions are formed under these stable ion conditions, the reactions could have also included an addition neighboring-halogen-elimination mechanism. Recently, Olah et al.[165] used cyanogen iodide or bromide in the presence of a slight excess of SbF_5 in SO_2 or SO_2ClF solution to prepare ethylenehalonium ions at a low temperature directly. For example, ethylene reacted with ICN-SbF_5 solution at $-60°$ to give the ethyleneiodonium ion. The reaction of ethylene with BrCN-SbF_5 in SO_2

$$CH_2{=}CH_2 \xrightarrow{ICN\text{-}SbF_5\text{-}SO_2} \overset{+}{I}\text{(three-membered ring)}$$

also resulted in the formation of polymeric products. However, ClCN-SbF_5 did not react with ethylene (or with 2,3-dimethyl-2-butene and adamantylideneadamantane) to give the corresponding chloronium ions, probably indicating the inverted polarity of cyanogen chloride compared with the other cyanogen halides. Tetramethylenebromo- and tetramethyleneiodonium ions have also been prepared by direct halogenation with the appropriate cyanogen halides and antimony pentafluoride.

7.5 Differentiation of π- and σ-Bonded Halogen Complexes of Alkenes

Strating, Wiering and Wynberg[166,167] reported that adamantylideneadamantane, a sterically highly hindered olefin, reacted with chlorine in hexane and bromine in CCl_4 solution to give the corresponding chloronium and bromonium salts. They proposed the structure of these rather insoluble salts (on which no spectroscopic study was conducted) to be that of three-membered-ring ethylenehalonium ions (σ complexes). Olah and coworkers[165] subsequently investigated these

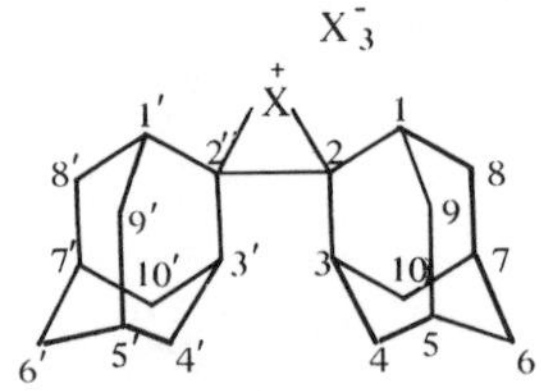

X_3^- $\overset{+}{X}$ 1′ 2′ 8′ 9′ 7′ 10′ 3′ 6′ 5′ 4′ 1 2 8 9 3 10 7 4 5 6

X = Cl, Br

stable salts by nmr (hydrogen-1 and carbon-13) spectroscopy, and concluded that they are not three-membered-ring halonium ions but molecularly bound π complexes.

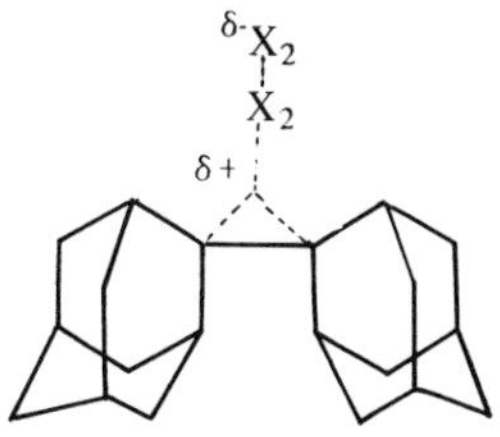

X = Cl, Br, I

Adamantylideneadamantane reacted with neat bromine or bromine in CCl_4, SO_2, HF-SO_2, or FSO_3H-SO_2 solution, or with bromine and $AgSbF_6$-SO_2, to give the same stable bromine π complex. Similar π complexes were also prepared by treating adamantylideneadamantane with *N*-bromosuccinimide or cyanogen bromide in FSO_3H-SO_2 solution. At $-30°$ adamantylideneadamantane reacted with FSO_3H-SO_2 to give the rapidly equilibrating carbenium ion which on treatment with bromine or chlorine forms the halogen π complexes. The

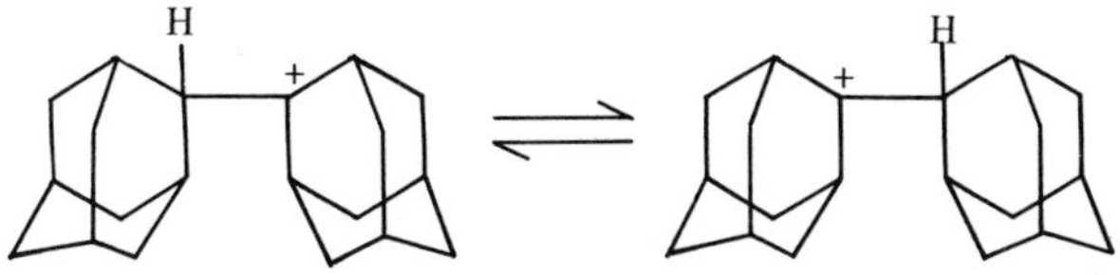

iodine complex was also obtained by adding adamantylideneadamantane to a solution of cyanogen iodide in FSO_3H-SO_2 at $-20°$. The pmr spectra of the halogen π complexes of adamantyleneadamantane are shown in Figure 12. The carbon-13 shifts are summarized in Table 37. The carbon shifts of C-2 and C-2′ are decisive in determining if the complexes are indeed three-membered-ring halonium ions (σ complexes), molecularly bonded π complexes, or rapidly equilibrating 2-halocarbenium ions.

Rapidly equilibrating 2-halocarbenium ions can be eliminated, because the chemical shift of the olefinic carbons C-2 and C-2′ at $\delta_{^{13}C}$ 36.0 (X = Cl), 158.7 (X = Br), and 153.7 (X = I) do not agree with the calculated average shift (e.g., approximately $\delta_{^{13}C}$ 204) for rapidly equilibrating systems. In addition, the steric restriction placed on the movement of the halogen atoms by the C-8 and C-10 and C-8′ and C-10′ protons also seems to prevent this halogen complex from being either a rapidly equilibrating ion pair, a mixture of bridged and equilibrating ions, or a rapidly equilibrating pair of partially bridged ions.

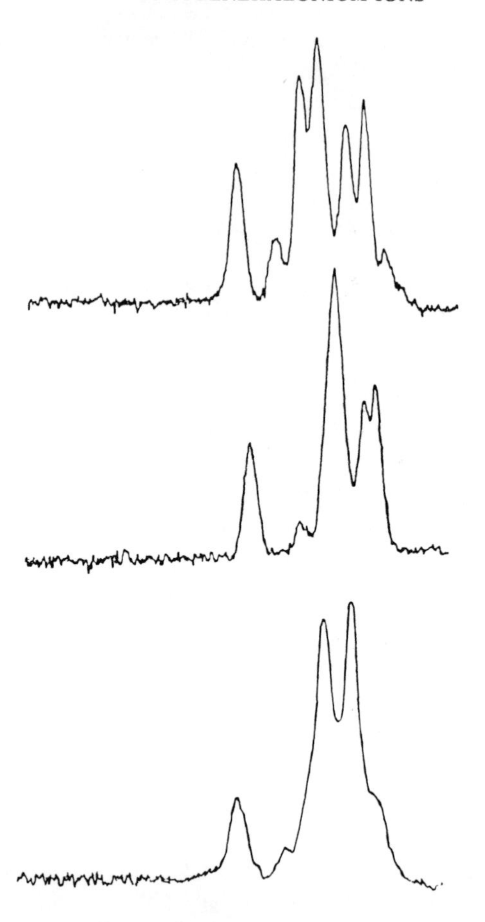

Figure 12. Pmr spectra of the adamantylideneadamatane-iodine complex in FSO_3H-SO_2 at $-20°$ (top), of the halogen complex in HF-SO_2 at $-30°$ (center), and of adamantylidene-adamantane in FSO_3H-SO_2 at $-30°$.

The halogen adducts therefore are most likely π-bonded halogen complexes. There is also additional evidence that these structures are different from three-membered-ring ethylenehalonium ions (σ complexes). The adamantylidenadamantane-bromine complex in CCl_4 transfers bromine irreversibly to cyclohexene or tetramethylethylene, forming the corresponding dibromides and adamantylideneadamantane. In addition, all other halogen adducts on quenching with nucleophiles also yield the precursor olefin, rather than any 1,2-addition products. Unlike ethylene, adamantylidenadamtane forms only π complexes with HF in SO_2ClF or with SbF_5 in SO_2ClF, which can be readily converted back, by the addition of nucleophiles, to the starting olefin. The

TABLE 37

Carbon-13 Chemical Shifts of Adamantylideneadamantane Complexes[a]

Compound	Solvent[b]	C-2	C-1 and C-3	C-4 and C-9 (C-8 and C-10)	C-8 and C-10 (C-4 and C-9)	C-5 and C-7	C-6
Adamantylidene-adamantane	C_6H_6	131.7	32.3	39.7	39.7	29.0	37.6
Protonation	SO_2 (−80°)	208.6	50.3	46.7	46.7	31.5	37.8
Bromine complex	Br_2	158.7	39.9	45.1	43.0	29.4	38.9
	SO_2ClF (−40°)	157.7	39.8	44.9	42.4	29.1	39.2
Chlorine complex	SO_2 (−70°)	157.8	37.7	43.4	39.7	27.4	36.7
Iodine complex	SO_2 (−70°)	153.7	38.5	44.0	43.4	28.0	37.8
SbF_5 complex	SO_2 (−82°)	156.7	36.8	42.4	38.8	26.6	35.8
	SO_2ClF (−80°)	157.3	39.6	43.6	37.6	27.4	36.6
HF complex	SO_2ClF (−80°)	155.1	36.0	41.6	38.1	26.0	35.2

[a]Carbon-13 shifts are referred to TMS in part per million.

[b]Probe temperature 30-35° unless otherwise indicated.

cmr shifts in these π complexes are very similar to those observed for the halogen complexes (Table 37). The observation of these π complexes is significant, since they seem to represent a clear differentiation of σ and π complexes in electrophilic additions to olefins.

CHAPTER 8

Tetramethylenehalonium Ions

8.1 PREPARATION VIA HALOGEN PARTICIPATION AND PMR STUDY

1,4-Halogen participation was first postulated to occur in the acetolysis of 4-iodo- and 4-bromo-1-butyl tosylates.[168] In subsequent studies Peterson and coworkers found anomalous rates in the addition of trifluoroacetic acid to 5-halo-1-hexenes[169,170] and 5-halo-1-pentynes.[171] Such observations were recognized as due to 1,4-halogen participation.[10] Recently, a study of the solvolysis of δ-chloro- and δ-fluoroalkyl tosylates indicated rate-accelerating 1,4-chlorine participation effects up to 99-fold.[172]

Direct experimental evidence for 1,4-halogen participation comes from the direct observation (by nmr spectroscopy) of five-membered-ring tetramethylenehalonium ions by Olah and Peterson,[15] and by Olah et al.[173] The pmr parameters of all the tetramethylehehalonium ions studied and directly observed are listed in Table 38.

8.1.1 Parent Tetramethylenehalonium Ions

The ionization of 1,4-dihalobutanes in SbF_5-SO_2 solution gave the parent tetramethylenehalonium ions.[15] The pmr spectra of these ions are similar

$$XCH_2CH_2CH_2CH_2X \xrightarrow{SbF_5\text{-}SO_2,\ -60^\circ} \text{(tetramethylene-}\overset{+}{X}\text{ ring)} \quad SbF_5X^-$$

$$X = Cl,\ Br,\ I$$

to each other, and one of them, the tetramethylenebromonium ion, is shown in Figure 13. Subsequently it was found that even 1,2- and 1,3-dihalobutanes

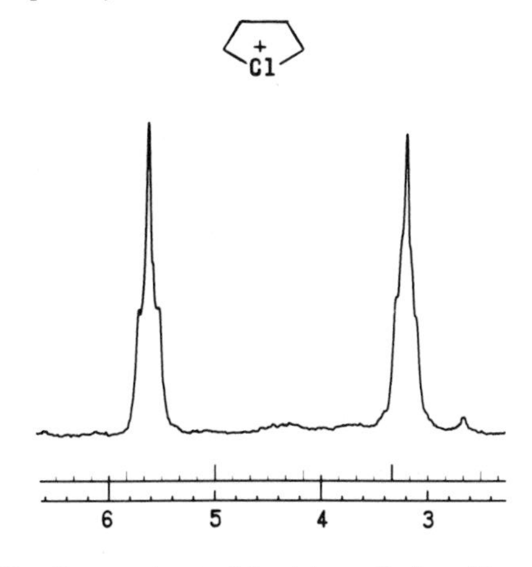

Figure 13. Pmr spectrum of the tetramethylenechloronium ion.

TABLE 38

Pmr Data of Tetramethylenehalonium Ions[a]

R_1	R_2	R_3	R_4	R_5	R_6	X	$\delta\ CH_2\overset{+}{X}$	$\delta\ CH_2C\overset{+}{X}$	$\delta\ CCHX^+$	$\delta\ CH_3$	δ CH
H	H	H	H	H	H	Cl	5.20	2.73			
H	H	H	H	H	H	Br	5.20	2.85			
H	H	H	H	H	H	I	5.00	2.80			
CH_3	H	H	H	H	H	Cl	5.20	2.85	6.60	2.10	
CH_3	H	H	H	H	H	Br	5.19	2.85	6.60	2.10	
CH_3	H	H	H	H	H	I	4.98	2.88	6.25	2.12	
H	H	H	CH_3	H	H	Cl	4.9-5.4	2.1-2.7		1.17	2.9-3.2
CH_3	H	CH_3	H	H	H	Cl		2.8	6.32	2.08	
H	CH_3	CH_3	H	H	H	Cl		2.8	6.32	2.08	
CH_3	H	CH_3	H	H	H	Br		2.9	6.43	2.15	
H	CH_3	CH_3	H	H	H	Br		2.9	6.43	2.16	
CH_3	H	CH_3	H	H	H	I		2.83	6.30	2.12	
H	CH_3	CH_3	H	H	H	I		2.83	6.30	2.22	
C_2H_5	H	H	H	H	H	Cl	5.1	2.1;2.8;3.1	6.4	1.20	
C_2H_5	H	H	H	H	H	Br	5.1	2.1;2.4;2.9	6.4	1.20	

CH_3	CH_3	CH_3	H	H	H	Cl		2.88-2.61	5.82	1.77 (d)	
										2.23 (s)	
CH_3	CH_3	CH_3	H	H	CH_3	Cl		2.87 (s)		2.13 (s)	
H	H	H	Cl	H	H	Cl	5.0-5.7	2.6-3.2			5.0-5.7
H	H	H	Br	H	H	Br	4.9-5.6	2.7-3.3			4.9-5.6
H	H	H	Cl	Cl	H	Cl	5.52-5.75				5.27
H	H	H	$(CH_2)_2$	$(CH_2)_2$	H	Cl	4.8-5.7				2.9-3.1

[a]From external capillary TMS in parts per million. Spectra were recorded at −60°. s, singlet; d, doublet.

when reacted with SbF_5-SO_2 or CH_3F-SbF_5-SO_2 solution[174] give the same tetramethylenehalonium ions.[175]

$XCH_2CH_2XCH_2CH_3$

$XCH_2CH_2CHXCH_3$

$XCH_2CH_2CH_2CH_2X$

$XCH_2CH_2CH_2CH_2I$

X = Br, Cl

$\xrightarrow{CH_3F\text{-}SbF_5\text{-}SO_2,\ -40^\circ}$ (tetramethylenehalonium ion, $\overset{+}{X}$ in ring) + $CH_3X^+CH_3$

X = Cl, Br, I

Reaction of 1-fluoro-4-iodobutane with SbF_5-SO_2 solution, however, gives, besides the tetramethyleneiodonium ion also the 1-butyleneidonium ion. The mechanism of the formation of the latter seems to involve ring opening and hydrogen shift, leading to the formation of the butyleneiodonium ion.

1,4-Difluorobutane, however, reacts with SbF_5-SO_2 or SbF_5-FSO_3H-SO_2 solution to give rearranged, open-chain fluorocarbenium ions, indicating the lack of bridging ability of fluorine. To our knowledge no fluoronium ions have thus far been identified in solution.

8.1.2 2-Methyltetramethylenehalonium Ions

Several alternate methods have been found for preparing 2-methyltetramethylenehalonium ions. For example the 2-methyltetramethylenechloronium ion can be obtained (1) by protonation of 5-chloro-1-pentene in SbF_5-FSO_3H-SO_2 solution at -60° (π route),[15] (2) by ionization of 5-chloro-2-pentyl trifluoroacetate with the same superacid system (σ route),[15] and (3) by ionization of 1,5-dichloropentane with SbF_5-SO_2.[176]

$CH_3CH(OCOCF_3)CH_2CH_2CH_2Cl$

or $CH_2{=}CHCH_2CH_2CH_2Cl$

$\xrightarrow{SbF_5\text{-}FSO_3H\text{-}SO_2,\ -60^\circ}$

$ClCH_2CH_2CH_2CH_2CH_2Cl$

$\xrightarrow{SbF_5\text{-}SO_2,\ -60^\circ}$

→ (2-methyltetramethylenechloronium ion, $\overset{+}{Cl}$ in ring)

The corresponding bromonium ion[15] can be prepared by (1) reaction of 1-bromo-4-chloropentane and (2) 1,5-dibromopentane with SbF_5-SO_2 solution at -60°. In the latter case, the six-membered-ring pentamethylenebromonium

$CH_3CHBrCH_2CH_2CH_2Br$

or

$BrCH_2CH_2CH_2CH_2CH_2Br$

$\xrightarrow{SbF_5\text{-}SO_2,\ -60^\circ}$ (2-methyltetramethylenebromonium ion, $\overset{+}{Br}$ in ring) SbF_5X^-

ion was not observed, or else was formed as an intermediate which rearranged to give the more stable 2-methyltetramethylenebronomium ion.

5-Iodo-1-pentene reacted with SbF_5-FSO_3H-SO_2 in a similar fashion to give the 2-methyltetramethyleneiodonium ion. The pmr spectral data of ions

$$CH_2{=}CHCH_2CH_2CH_2I \xrightarrow{SbF_5\text{-}SO_2,\ -60^\circ}$$

(X = Cl, Br, I) are listed in Table 38.

Recently, Peterson and Bonazza[177] reported the preparation of the 3-methyltetramethylenechloronium ion from 1,4-dichloro-2-methylbutane in SbF_5-SO_2 solution at -78°. The dimethyl-β-chloroethylcarbenium ions (32%) was also formed in the reaction (68%). The structure of the ions was based on their pmr

$$CH_3CH(CH_2Cl)CH_2CH_2Cl \xrightarrow{SbF_5\text{-}SO_2} \text{(ring, Cl}^+\text{)} + (CH_3)_2\overset{+}{C}CH_2CH_2Cl$$

68% 32%

$\downarrow$ -23°, 10 min

$$\text{(ring, Cl}^+\text{)} + (CH_3)_2\overset{+}{C}CH_2CH_2Cl$$

55% 45%

parameters. When the solution containing the mixture is warmed at -23° for 10 min, the 3-methyltetramethylenechloronium ion undergoes an apparent 1,2-dimethyl shift to give the 2-methyl-substituted ion. The increase in the relative amount of the dimethyl-β-chloroethylcarbenium ion indicates that the 3-methyltetramethylenechloronium ion on warming gives the hydrogen-shifted product (dimethyl-β-chloroethylcarbenium ion) along with the methyl-shifted product (2-methyltetramethylenechlonenium ion).

8.1.3 2,5-Dimethyletramethylenehalonium Ions

The π route of preparation through protonation of 5-halo-1-hexenes was used for the preparation of 2,5-dimethyltetramethylenehalonium ions.[15] The

$$CH_2{=}CHCH_2CH_2CHXCH_3 \xrightarrow{SbF_5\text{-}FSO_3H\text{-}SO_2,\ -60^\circ}$$

X = Cl, Br, I

formation of cis and trans isomers was clearly shown by the appearance of a pair of methyl doublets in the case of the 2,5-dimethyltetramethyleneiodonium ion (Figure 14).

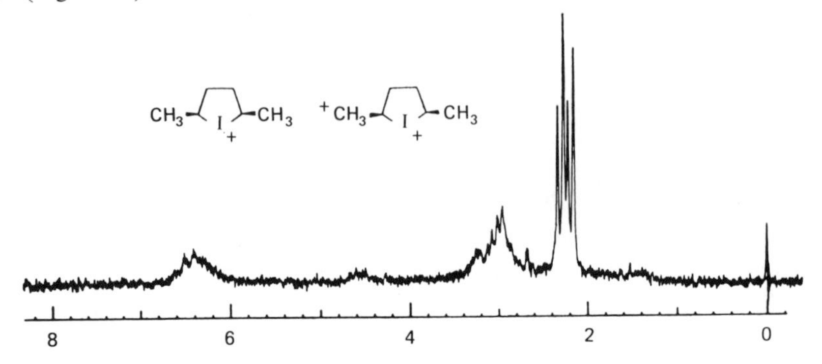

Figure 14. Nmr spectrum of *cis*- and *trans*-2,5-dimethyltetramethyleneiodonium ions.

The isomeric (cis and trans) 2,5-dimethyltetramethylenebromonium ions were also obtained (as an equimolar mixture) when 2,5-dibromohexanes were treated with SbF_5-SO_2 solution at $-70°$.[169]

2,2-Dimethyltetramethylenehalonium ions can also be prepared in several ways. Thus the 2,2-dimethyltetramethylenechloronium ion was obtained by ionization of (1) 2,5-dichloro-2-methylpentane in SbF_5-SO_2 or SbF_5-FSO_3H-SO_2, (2) 2-methyl-5-chloro-2-pentanol in SbF_5-FSO_3H-SO_2, (3) 4-chloro-1-fluoro-4-methylpentane[173] in SbF_5-FSO_3H-SO_2, and (4) dichlorohexanes[176] in SbF_5-SO_2 .

$(CH_3)_2CClCH_2CH_2CH_2Cl$ — SbF_5-SO_2, $-60°$ / SbF_5-FSO_3H-SO_2

$(CH_3)_2CClCH_2CH_2CH_2Y$ — SbF_5-FSO_3H-SO_2, $-60°$

X = OH; Y = Cl

X = Cl; Y = F

$ClCH_2CH_2CH_2CH_2CH_2CH_2Cl$

$ClCH_2CH_2CH_2CH_2CHClCH_3$ — SbF_5-SO_2

$ClCH_2CH_2CH_2CHClCH_2CH_3$

→ 2,2-dimethyltetramethylenechloronium ion ($\overset{+}{Cl}$)

The mechanism of the formation of the 2,2-dimethyltetramethylenechloronium ion from 4-chloro-1-fluoro-4-methylpentane was postulated to involve a fluoronium ion as an unstable intermediate (although without involvement of a fluoronium ion an equally reasonable mechanism could also be suggested).

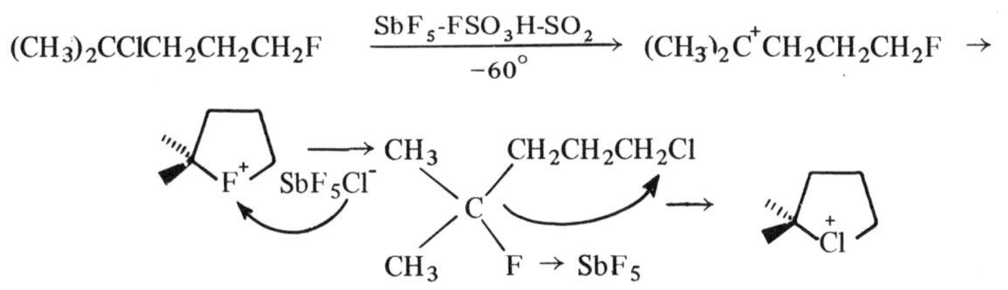

$$(CH_3)_2CClCH_2CH_2CH_2F \xrightarrow[-60^\circ]{SbF_5\text{-}FSO_3H\text{-}SO_2} (CH_3)_2\overset{+}{C}CH_2CH_2CH_2F \rightarrow$$

In the case of dichlorohexanes an intermediate was observed prior to the formation of the 2,2-dimethyltetramethylenehalonium ion.[26d,178] Ionization of dihalohexanes in SbF_5-SO_2 solution at -78° gave 2-ethyltetramethylenehalonium ions. The pmr spectra of these ions are consistent with their structure (Table

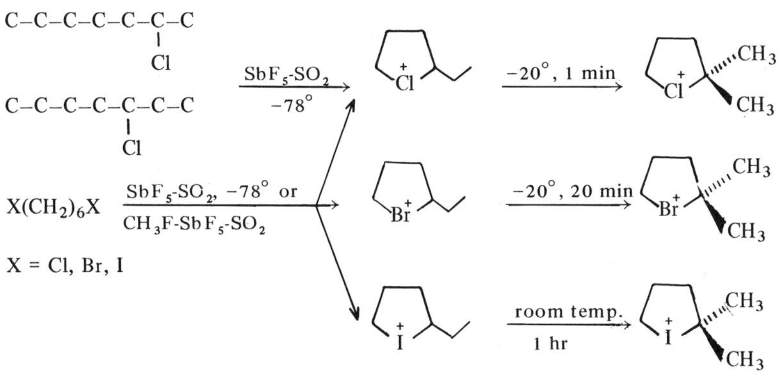

38). Both the 2-butyltetramethylenedibromonium and -bronomium ions can be transformed into the corresponding 2,2-dimethyl-substituted ions (via ring opening followed by 1,2-hydrogen and 1,2-methyl shifts) when their solutions are warmed to -20° for up to 20 min. Formation of the 2,2-dimethyltetramethyleneiodonium ion from the corresponding 2-ethyl-substituted iodonium ion was not observed even when the sample was allowed to stand at room temperature for 1 hr, although under these conditions significant decomposition occurred. The pmr spectrum of the 2,2-dimethyltetramethylenechloronium ion is shown in Figure 15.

2,2-Dimethytetramethylenebromonium and -iodonium ions could also be prepared from 2-methyl-5-halopentanol with SbF_5-FSO_3H-SO_2, or by protonation of 5-halo-2-methyl-2-pentene in "magic acid," SbF_5-FSO_3H-SO_2.

Figure 15. Pmr spectrum of the 2,2-dimethyltetramethylenechloronium ion.

$$(CH_3)_2COH(CH_2)_3X \xrightarrow[-60^\circ]{SbF_5\text{-}FSO_3H\text{-}SO_2}$$

X = Br, I

$$(CH_3)_2C{=}CH(CH_2)_2X \xrightarrow{SbF_5\text{-}FSO_3H\text{-}SO_2,\ -78^\circ}$$

(both give the 2,2-dimethyltetramethylenehalonium ion, $\overset{+}{X}$ in ring) X = Br, I

Attempts were made to prepare the 2,2-dimethyltetramethylenefluoronium ion by ionizing the potential difluoride precursors in SbF_5-SO_2, but only open-chain tertiary ions were formed.[173] The lack of aliphatic fluorine resonance absorption of J_{HF} coupling indicates that the fluorine atom is complexed

$$(CH_3)_2CF(CH_2)_3F \xrightarrow{SbF_5\text{-}SO_2,\ -78^\circ} (CH_3)_2\overset{+}{C}-(CH_2)_3F{\rightarrow}SbF_5 \quad SbF_6^-$$

and rapidly exchanges with SbF_5 (a similar rapidly exchanging donor-acceptor complex of methyl fluoride with SbF_5 has been reported[179]).

The pmr data of the 2,2-dimethyltetramethylenehalonium ions are summarized in Table 38.

8.1.4 Tri- and Tetramethyltetramethylenehalonium Ions

Peterson and Henrichs[178] reported the preparation of 2,2,5-trimethyl- and 2,2,5,5-tetramethyltetramethylenechloronium ions from 2,5-dichloro-2-methylhexane and 2,5-dichloro-2,5-dimethylhexane, respectively. The structure of these ions was studied by both proton and carbon-13 nmr spectroscopy (Tables 38 and 39).

C-C(C)(Cl)-C-C-C(R)(Cl)-C $\xrightarrow{SbF_5\text{-}SO_2,\ -78^\circ}$ [cyclic chloronium ion, Cl^+, R]

R = H, CH_3 R = H; R = CH_3

It is of interest to note the observation of only one proton and carbon-13 absorption line for the nonequivalent geminal methyl groups in the dimethyltetramethylenechloronium ion. These data, as well as the modest temperature dependence of the carbon shifts, suggests that an equilibrium exists with the open-chain carbenium ion causing the seeming equivalence of the methyl groups.

It would be expected that the tetramethyltetramethylenechloronium ion could be prepared from 2,5-dichloro-2,5-dimethylhexane, but an earlier study indicated that an open-chain dicarbenium ion is instead obtained. It was found, however, that the halonium ion could be obtained when additional dichloride was added to the solution of the dication. A possible reaction pathway is

[dication] + [dichloride (Cl, Cl)] $\rightleftharpoons$ [open-chain cation] + [chloronium ion, Cl^+, Cl] $\rightleftharpoons$ 2 [chloronium ion, Cl^+]

Attempts to prepare 2-hydroxy-2-methyltetramethylenehalonium ions by protonation of methyl γ-halopropyl ketones in superacids were unsuccessful.[178] Instead, the open-chain ions were formed with no halogen participation.

[methyl γ-halopropyl ketone (O, X)] $\xrightarrow{SbF_5\text{-}FSO_3H\text{-}SO_2}$ [H–O^+ protonated open-chain ketone, X]; ✕→ [cyclic halonium ion: HO, CH_3, X^+]

8.1.5 Halogenated Tetramethylenehalonium Ions

Peterson and coworkers[176] reported the preparation of 2-halomethyltetramethylenehalonium ions (X = Cl, Br) by the ionization of 1,2,5-trihalopentanes

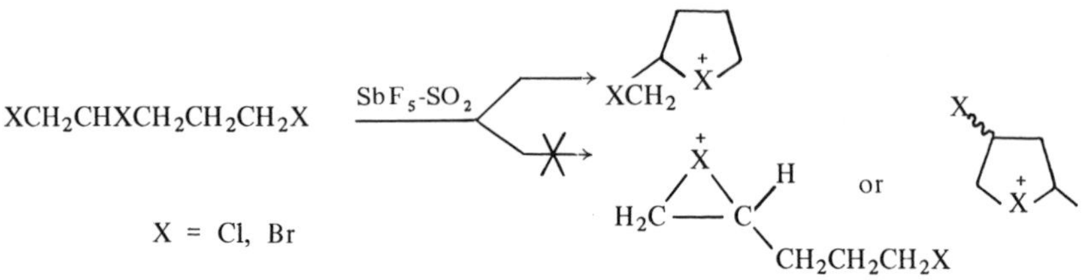
XCH2CHXCH2CH2CH2X
SbF5-SO2
XCH2
X
H2C
C
H
CH2CH2CH2X
or
X = Cl, Br

in SbF_5-SO_2 solution. As expected, ionization took place at the secondary rather than the primary halide centers. Isomeric halonium ions were not formed under the experimental conditions.

When 1,2,4-trichloro- and 1,2,4-tribromobutane were allowed to react with SbF_5-SO_2 at $-60°$, the 3-chlorotetramethylenechloronium ion and the 3-bromotetramethylenebromonium ion, respectively, were obtained.[177] The pmr data are tabulated in Table 39.

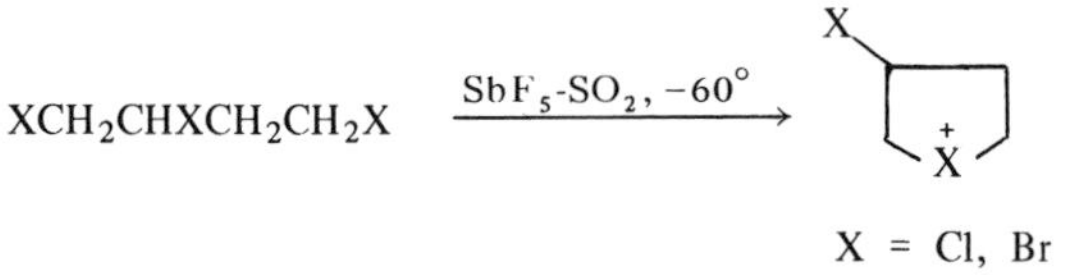
XCH2CHXCH2CH2X
SbF5-SO2, −60°
X
X = Cl, Br

Similarly, when *meso*-1,2,3,4-tetrachlorobutane was ionized in SbF_5-SO_2 solution at $-23°$ for approximately 1.5 hr, or at 0° for 1 min, the *cis*-4,5-dichlorotetramethylenechloronium ion was formed.[177]

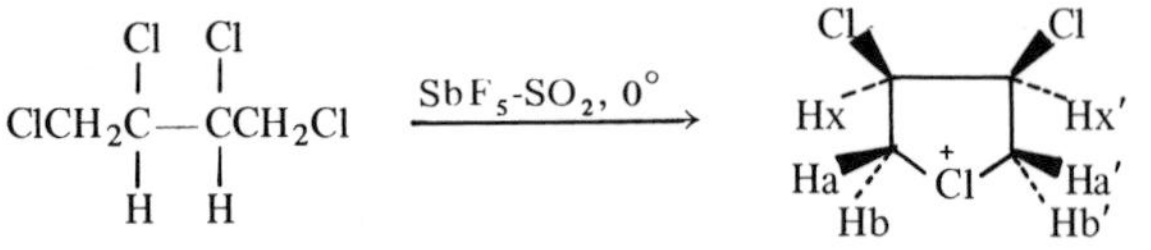
Cl Cl
ClCH2C—CCH2Cl
H H
SbF5-SO2, 0°
Cl
Cl
Hx
Hx'
Ha
Ha'
Cl
Hb
Hb'

8.1.6 2-Methylenetetramethylenehalonium Ions

Peterson and coworkers[171] obtained evidence for halogen participation in the addition of trifluoroacetic acid to 5-halo-1-pentynes. Olah and coworkers

$$HC{\equiv}C(CH_2)_3X \xrightarrow{CF_3COOOH} HC(Cl){=}CH(CH_2)_3OOCCF_3$$

subsequently examined the protonation of 5-halo-1-pentynes in superacids in order to see whether the corresponding 2-methylenetetramethylenehalonium ions could be observed as stable species in superacid solutions. The results

$$HC{\equiv}C(CH_2)_3I \xrightarrow{SbF_5\text{-}FSO_3H\text{-}SO_2,\ -78^\circ}$$

δ3.81 δ3.25 δ7.37 H I⁺ δ5.71 H δ9.85

showed that 5-iodo-1-pentyne was smoothly protonated to give the corresponding 2-vinyltetramethyleneiodonium ion. However, protonation of 5-chloro- or 5-fluoro-1-pentyne in SbF_5-FSO_3H-SO_2 gave only unidentifiable products.

8.2 CARBON-13 NMR STUDY OF TETRAMETHYLENEHALONIUM IONS

The carbon-13 nmr of tetramethylenehalonium ions has been studied by Olah[26d] and Peterson[180] and their coworkers. Carbon-13 chemical shift data for tetramethylenehalonium ions are shown in Table 39. The C-3 and C-4 resonances in unsymmetrically substituted tetramethylenehalonium ions were differentiated by assuming that C-2 had greater electron deficiency than C-5 and thus exerted a slightly larger deshielding α-substituent effect. The α-substituent effect of a positively charged carbon has been observed to be that of deshielding in a large number of carbenium ions.

The ions in Table 39 can be divided into symmetrically and unsymmetrically substituted ions.

8.2.1 Symmetrically Substituted Ions

Whereas the approach of observing α- and β-$^+$XR substituent effects to estimate ring carbon shifts in ethylenehalonium ions was unsuccessful, good results were obtained for five-membered-ring halonium ions. For example, the methyl shift in *n*-butylalkylchloronium ions should be very similar to the methyl shift in *n*-butane ($\delta_{^{13}C}$ 13.3), since the δ-substituent effect for $^+$XR should be negligible. Cyclization of the CH_3–$CH_2CH_2CH_2Cl^+$ group to the tetramethylenechloronium ion should result in a downfield shift of 67.5 ppm in the methyl carbon shielding. The estimated shift of $\delta_{^{13}C}$ 80.8 for C-2 and C.5 in the tetramethylenechloronium ion is in good agreement with the experi-

TABLE 39

Carbon-13 Nmr Shifts of Tetramethylenehalonium Ions[a,26d,180]

Halonium ion	Substituent	Temp. (°C)	C-2	C-3	C-4	C-5	CH_2	CH_3
Cl[b]		-40°	77.9	33.9				
Br[b]		-40°	70.8	36.4				
I[b]		-40°	51.1	38.8				
Cl[c]	2-CH_3	-67.8°	113.6	41.5	35.5	75.4		22.6
		-54.3°	113.8	41.5	35.5	75.4		22.6
		-36.0°	113.8	41.5	35.3	75.3		22.6
		-10.0°	115.6	42.9	36.7	76.7		22.9
Br	2-CH_3	-40°	107.0 (106.7)	44.8 (44.3)	37.5 (36.9)	69.8 (69.7)		24.2 (23.7)
I[d]	2-CH_3	-55°	83.6	41.8	38.4	49.4		24.9
Cl	3-CH_3	-55°	80.0	43.0	40.6	78.4		25.1
Br (cis)	2,5-$(CH_3)_2$	-40°	97.5	39.1				19.3
Br (trans)	2,5-$(CH_3)_2$	-40°	98.9	40.9				19.8
Cl[e]	2,2-$(CH_3)_2$	-69.3°	192.4	49.0	37.1	66.4		35.3
		-56.1°	198.4	49.5	37.0	65.5		35.7
		-41.3°	204.8	50.0	37.0	64.5		36.2
		-40°	198.2	52.5	37.8	63.9		38.7
Br	2,2-$(CH_3)_2$	-69.3°	151.1	50.2	37.8	66.2		33.9
		-56.1°	152.0	50.2	37.8	66.0		33.9
		-41.3°	153.1	50.3	37.8	65.8		34.0
		-40°	195.8	50.3	37.7	65.5		33.9
Cl	2-C_2H_5	-65°	121.5	39.3	35.4[e]	74.4	23.9[e]	13.8
Br	2-C_2H_5	-55°	117.2	42.7	37.6	69.1	32.7	16.0
		-65°[f]	115.9	41.8	36.9[e]	68.5	31.7[e]	15.4
I	2-C_2H_5	-65°	95.5	45.3[e]	38.6	40.7	32.8[e]	18.0

Cl	2,2,5-$(CH_3)_3$	−67.8°	154.8	47.6	43.6	99.5	33.1	22.6
		−54.3°	154.8	47.6	43.5	99.0	33.1	22.5
		−36.0°	160.7	47.8	43.6	98.4	33.4	22.6
Cl	2,2,5,5-$(CH_3)_4$	−67°	137.3	47.6	47.6	137.3	32.2	
		−56°	137.6	47.7	47.7	137.6	32.2	
Cl	3-Cl	−55°	81.3	62.0	42.5	77.0		
Br	3-Br	−55°	74.4	52.4	44.9	69.5		
Cl (cis)	4,5-Cl_2	−55°	86.0	64.9				

[a] Carbon shifts are referred to TMS in SO_2 solution in parts per million.
[b] For symmetric ions C-2 and C-3 are equivalent to C-5 and C-4, respectively.
[d] Assignment of C-3 and C-4 may be reversed.
[e] Assignments are tentative.

mentally determined value of $\delta_{^{13}C}$ 77.9. This procedure is also successful in estimating the C-3 and C-4 shift in the ion, as well as the C-2 and C-5 and C-3 and C-4 shifts in related bromonium and iodonium ions.

α- and β-substituent effects for the RX^+ group in the parent tetramethylenehalonium ions are therefore the same as in dialkylhalonium ions. Since the equilibrium positions with open-chain halocarbenium ions are far to the side of bridged ions for primary systems, it follows that the parent tetramethylenehalonium ions all exist as static bridged species, with no rapid equilibration with open-chain ω-haloalkylcarbenium ions being involved.

The cmr and pmr spectra of the 1,4-dimethyltetramethylenebromonium ion reveal the presence of cis and trans isomers in equal amounts, also indicating that the ion is a symmetrically bridged ion. If it is in equilibrium with an open-chain form, it is a slow equilibrium with an undetectable (by nmr) amount of the open-chain ω-haloalkylcarbenium ion. The larger than usual methyl substituent effect (28 ppm) probably indicates an increased carbenium ion character for the C-2 and C-5 carbons.

The methyl carbon absorbance of the cis isomer is shielded by 2.5 ppm from that of the trans isomer, reflecting the larger steric interaction between the methyl groups in the former case. The effect is smaller than in *cis*- and *trans*-1,2-dimethylethylenebromonium ions, indicating that the methyl groups are closer in the cis isomer of this ion than in the *cis*-1,4-dimethyltetramethylenebromonium ion.

8.2.2 Unsymmetrically Substituted Ions

Replacement of a proton on C-2 by a methyl group in tetramethylenehalonium ions results in a deshielding of 32-38 ppm in the C-2 absorbance. This may indicate an unsymmetric bridged ion, or be a result of these ions being in equilibrium with open-chain ω-haloalkylcarbenium ions. However, the latter possibility is unlikely, as the methyl substituent effect is approximately the same for chlorine, bromine, and iodine, so that the equilibrium constants would have to be the same in all three cases.

Furthermore, a consideration of the cmr spectrum of 2-methyltetramethylenechloronium ions at different temperatures showed that the difference between the C-2 and C-5 chemical shifts changes only 0.7 ppm on lowering the temperature from -20 to $-90°$, while the corresponding change in $\delta_{C\text{-}2} - \delta_{C\text{-}3}$ is only 0.10 ppm.

Large downfield shifts at C-2 are observed on the C-2 substitution of a second methyl group into 2-methyltetramethylenechloronium and -bromonium ions. In the former case the α-substituent effect is 82 ppm, and in the latter 46 ppm. These large changes most likely indicate increasing carbenium ion character at C-2 with increasing methyl substitution. This is also suggested by the comparatively large downfield methyl shifts in 2,2-dimethyltetramethylenechloronium and -bromonium ions (38.7 and 33.9 ppm, respectively). The cmr spectra of the ions have also been recorded at different temperatures by Henrichs and Peterson[78], who found that the 1,1-dimethyltetramethylenechloronium ion is in rapid equilibrium with substantial amounts of the open-chain ω-chloroalkylcarbenium ion. They also found that the 1,1,4-trimethyltetramethylenecarbenium ion and 1,1-dimethyltetramethylenebromonium ion are in equilibrium with smaller amounts of the open-chain ion.

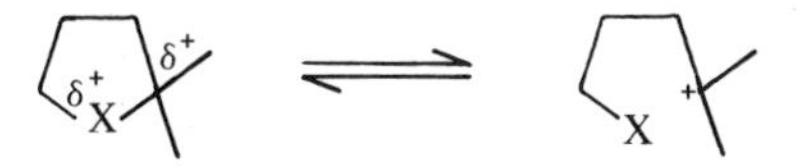

In contrast to the deshielding of the C-2 shift in less substituted ions, the C-5 shift becomes shielded, relative to the unsubstituted ion, with increasing methyl substitution. These effects are relatively small for monomethyl-, but more substantial in 2,2-dimethyl-substituted ions. For the methyldimethyltetramethylenechloronium ion the shift is further shielded (14 ppm) than in the corresponding -bromonium ion (5.3 ppm), because of the greater contribution of the equilibrating open-chain form in the former case. This is quite reasonable, considering the greater stability of a positively charged bromine atom as compared with a positively charged chlorine atom.

CHAPTER 9

Tri- and Pentamethylenehalonium Ions

9.1 TRIMETHYLENEHALONIUM IONS

Attempts to prepare trimethylenehalonium ions by ionizing the appropriate 1,3-dihalopropanes with SbF_5-SO_2 or with 1:1 FSO_3H-SbF_5-SO_2 solution at low temperature were unsuccessful.[174] The ions obtained were either three- or five-membered-ring halonium ions, formed through ring contraction or expansion, respectively. For example, when 1-halo-3-iodopropanes were reacted with SbF_5-SO_2ClF solution at $-78°$, the propyleneiodonium ion was formed. This ion could be formed through the trimethyleneiodonium ion which was, however, not observed as an intermediate. Alternatively, the nonassisted primary ion

$$I(CH_2)_3X \xrightarrow{SbF_5\text{-}SO_2ClF,\ -78°} [\text{trimethyleneiodonium ion}] \longrightarrow \text{propyleneiodonium ion } (CH_3, \overset{+}{I})$$

$$X = F, Cl, I$$

$$I(CH_2)_3X \longrightarrow (ICH_2CH_2CH_2^+) \longrightarrow ICH_2C^+HCH_3 \longrightarrow \text{propyleneiodonium ion}$$

could undergo rapid 1,2-hydrogen shift to the secondary ion, which then would form the propyleneiodoium ion via iodine participation.

Similarly, when 1-halo-3-bromopropanes were treated in the same way at $-78°$, the propylenebromonium ion was obtained. However, when dichloro-

$$Br(CH_2)_3X \xrightarrow[-78°]{SbF_5\text{-}SO_2 \text{ or } FSO_3H\text{-}SbF_5\text{-}SO_2} \text{propylenebromonium ion } (CH_3, \overset{+}{Br})$$

$$X = Cl, Br$$

propanes were treated with SbF_5-SO_2ClF solution at $-60°$, a rapidly equilibrating system was formed (see Section 7.1.2).

When 1-iodo-3-halobutanes were treated wtih SbF_5-SO_2 solution at $-60°$, three-membered-ring halonium ions were formed. In contrast, 1,2- and 1,3-dibromo(or chloro)butanes gave tetramethylenehalonium ions.

The ionization of 1,3-dihalo-2-methylpropanes with SbF_5-SO_2 solution gave both three- and five-membered-ring halonium ions and open-chain halocarbenium ions.

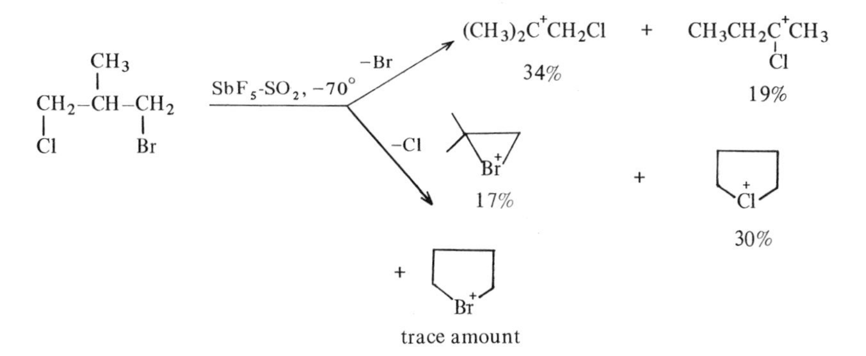

+ [four-membered ring Br^+]

trace amount

Another possible route to four-membered-ring halonium ions could be though dimethyl-β-haloethylcarbenium ions. These ions can be prepared from dihalopentanes with SbF_5-FSO_3-SO_2 solution at −78°. The proton and carbon-13 nmr data of the systems, however, indicated the absence of halogen participation.[26d]

$XCH_2CH_2C(CH_3)_2Y$

X = Cl; Y = OH
X = Br; Y = OH
X = Y = Br
or
$(CH_3)_2C(CH_2Br)_2$

$\xrightarrow{FSO_3H\text{-}SbF_5\text{-}SO_2,\ -78^\circ}$ $XCH_2CH_2C^+(CH_3)_2$
X = Cl, Br

The only reported preparation of long-lived four-membered-ring halonium ions is that of Exner et al.[18] 3,3-Bis(halomethyl)trimethylenebromonium ions were prepared by treating tetrahaloneopentanes with SbF_5-SO_2ClF solution at low temperature. Seemingly, halogen substitution stabilizes the four-membered-ring compared to ring contraction or expansion. The pmr spectrum of fluorinated ions shows a singlet at δ 5.28 and a doublet at δ 4.68 (J_{HF} = 47

$RCH_2C(CH_2R')_2CH_2Br$ $\xrightarrow{SbF_5\text{-}SO_2ClF,\ -55^\circ}$ $(RCH_2)_2C(CH_2)_2Br^+$

R = R′ = F
R = R′ = Br
R = F, R′ = Br

R = F
B = Br

Hz). In contrast, the brominated ion displays a temperature-dependent proton absorption at δ 5.17 (from internal $Me_4\overset{+}{N}\ BF_4^-$), indicating that the following exchange process occurs.

$$(BrCH_2)_2C(CH_2^+)(CH_2Br) \rightleftharpoons (BrCH_2)_2C(CH_2)_2Br^+ \rightleftharpoons (BrCH_2)(CH_2Br)C(CH_2)_2Br^+ \rightleftharpoons \text{, etc.}$$

9.2 PENTAMETHYLENEHALONIUM IONS

As discussed, attempts to prepare six-membered-ring halonium ions by treating 1,5-dihalopentanes with SbF_5-SO_2 gave exclusive rearrangement to five-membered-ring halonium ions. Recently, however, Peterson et al.[16] prepared

$$X(CH_2)_5X \xrightarrow{SbF_5\text{-}SO_2} \text{2-methyltetramethylenehalonium ion } (X^+)$$

X = Cl, Br (starting material); X = Cl, X = Br (product)

six-membered-ring halonium ions by the methylation of 1,5-dihalopentanes with methyl fluoroantimonate (CH_3F-SbF_5) in SO_2 solution. Some rearrangement to five-membered-ring halonium ions was also observed.

$$X(CH_2)_5X \xrightarrow{CH_3F\text{-}SbF_5\text{-}SO_2} \text{pentamethylenehalonium } (X^+) + \text{2-methyltetramethylenehalonium } (X^+)$$

X = Br (72-55%) X = Br (28-45%)

X = I (95%) X = I (5%)

Alternatively, six-membered-ring halonium ions were also formed when equimolar methyl fluoroantimonate was added to dihalonium ions.

The dihalonium ions were prepared from 1,5-dihalopentane and 2 moles of

$$X(CH_2)_5X \xrightarrow{2CH_3F\text{-}SbF_5\text{-}SO_2} 2\ CH_3X^+(CH_2)_5X^+CH_3\ SbF_6^- \xrightarrow{X(CH_2)_5X} 2\ \text{pentamethylenehalonium } (X^+)$$

X = Br, I X = Br, I

methyl fluoroantimonate. Furthermore, the dimethylbromonium ion is also a sufficiently active methylating agent to form cyclic pentamethylenebromonium ions from 1,5-dibromopentane.

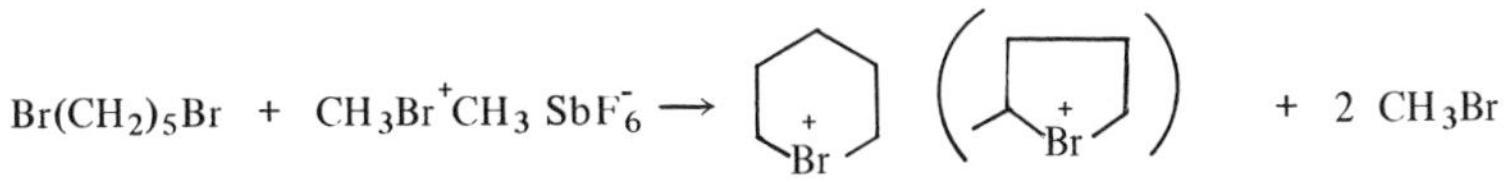

$$Br(CH_2)_5Br + CH_3Br^+CH_3\ SbF_6^- \longrightarrow \text{(pentamethylenebromonium ion)}\ \left(\text{2-methyltetramethylenebromonium ion}\right) + 2\ CH_3Br$$

73% 27%

The proton and carbon-13 shifts (the latter shown in parentheses) of pentamethylenehalonium ions are consistent with their structures. The penta-

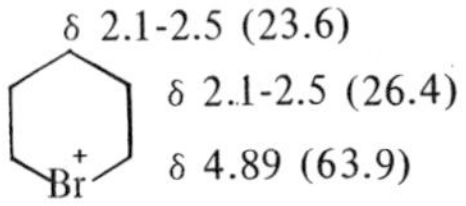

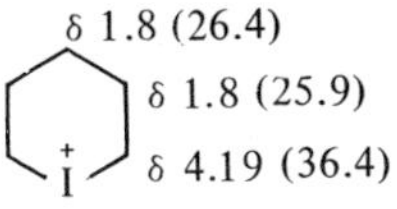

Pentamethylenebromonium ion (Br^+): δ 2.1-2.5 (23.6); δ 2.1-2.5 (26.4); δ 4.89 (63.9)

Pentamethyleneiodonium ion (I^+): δ 1.8 (26.4); δ 1.8 (25.9); δ 4.19 (36.4)

methyleneiodonium ion rearranged in part to 2-methyltetramethyleneiodonium ions after 2 hr at room temperature in SO_2 solution. The pentamethylenebromonium ion, however, rearranged smoothly to the 2-methyltetramethylenebromium ion (50%) over a period of 20 min at $-20°$ (complete rearrangement requires a longer period of time). Quenching of a mixture of the iodonium ions with methanol gave a 60% isolated yield of the corresponding ω-iodoalkyl ethers, $I(CH_2)_5OCH_3$ and $I(CH_2)_3CH(OCH_3)CH_3$.

CHAPTER 10

Bicyclic Halonium Ions

10.1 CYCLOPENTENEBROMONIUM ION (2-BROMONIA[3.1.0] BICYCLOHEXANE)

Although halogen additions to cycloalkenes are also assumed to proceed through the corresponding bicyclic halonium ions, these ions are quite elusive. Olah, Liang and Staral[23] recently prepared the cyclopentenebromonium ion by the ionization of *trans*-1,2-dibromocyclopentane in SbF_5-SO_2ClF solution at $-120°$. The pmr (60 MHz) spectrum of the ion solution showed a broadened peak at δ 7.32

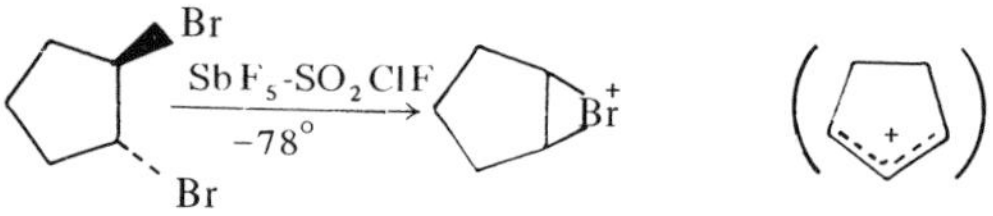

$$\xrightarrow[-78°]{SbF_5\text{-}SO_2ClF}$$

(two protons) and two broad peaks centered at δ 3.14 (four protons) and δ 2.50 (two protons). The nmr spectra of the solution also showed the presence of the related cyclopentenyl cation. When the solution was slowly warmed to $-80°$, the cyclopentenebromonium ion gradually and cleanly transformed into the allylic ion. The cyclopentenyl ion present initially in solution might be formed as a result of local overheating during preparation.

The cyclopentenebromonium ion was also obtained via protonation of 4-bromocyclopentene in FSO_3H-SbF_5-SO_2ClF solution at $-120°$, through the following reaction sequence.

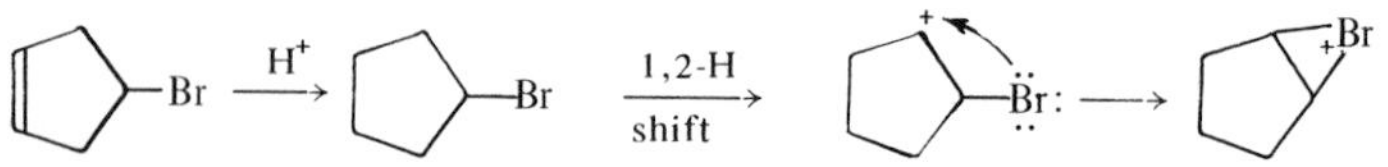

$$\xrightarrow{H^+} \quad \xrightarrow[\text{shift}]{1,2\text{-H}} \quad \longrightarrow$$

The proton noise-decoupled carbon-13 spectrum of the cyclopentenebromonium ion (Figure 16) shows three carbon resonances at $\delta_{^{13}C}$ 114.6 (doublet, J_{CH} 190.6 Hz), 31.8 (triplet, J_{CH} 137.6 Hz), and 18.7 (triplet, J_{CH} 134.0 Hz). In addition, the cmr spectrum shows three additional minor peaks which exactly correspond to those reported for the cyclopentenyl cation [The cmr assignments were made with the aid of proton-coupled cmr spectra (Figure 16).] The bicyclic structure of the cyclopentenebromonium ion is therefore firmly established.

It is instructive to compare both proton and carbon shifts of the cyclopentene-

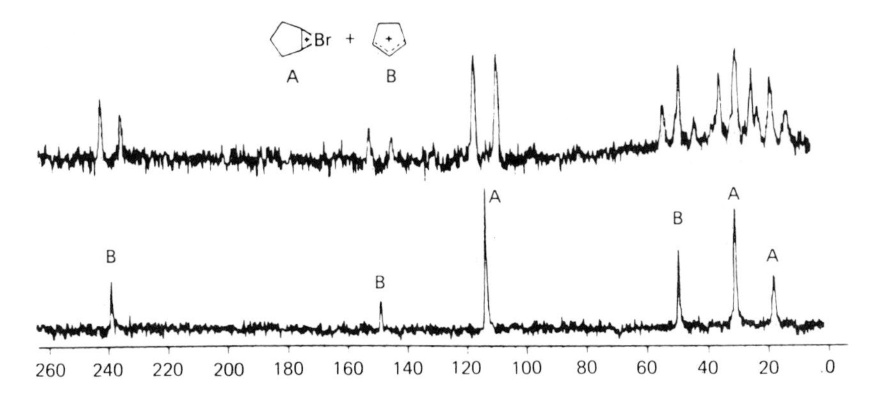

Figure 16. (*a*) Proton noise-decoupled carbon-13 nmr spectrum of 1 (peaks marked *A*) and 3 (peaks marked *B*). (*b*) Proton-coupled carbon-13 nmr spectrum.

bromonium ion with those of suitable model ions. The most closely related known cyclic bromonium ion containing a three-membered ring is the *cis*-1,2-dimethylethylenebromonium ion.

The chemical shift for the two carbon atoms where bridging takes place is found at $\delta_{^{13}C}$ 110.9, which is in good agreement with that in cyclopentene-

H Br H 114.6 7.32

CH_3 CH_3 H Br H 110.9 6.90

bromonium ion (114.6). The proton shifts for the two equivalent olefin-type protons in both ions are also similar.

Attempts were also made to prepare the cyclopentenechloronium ion via ionization of *trans*-1,2-dichlorocyclopentane in SbF_5-SO_2ClF at −120°. However, instead of the cyclopentenechloronium ion, only the 1-chloro-1-cyclopentyl cation was obtained. Apparently, the participation of the smaller chlorine atom could not effectively compete with the fast 1,2-hydride shift forming the

Cl Cl +Cl + —Cl

tertiary ion. The larger bromine atom, however, preferentially participates with the neighboring electron-deficient center, forming the bicyclic bridged ion.

10.2 3-CHLORONIA[4.3.0]BICYCLONONANE

Peterson and Bonazza[22] reported that ionization of *cis*-1,2-bis(chloromethyl)-cyclohexane in SbF_5-SO_2 solution at $-78°$ gives the bicyclic five-membered-ring chloronium ion along with smaller amounts of other species. Warming the solution containing the ion to $-10°$ leads to the formation of the open chain tertiary carbenium ion.

$$\text{cis-1,2-}(CH_2Cl)_2C_6H_{10} \xrightarrow{SbF_5\text{-}SO_2,\ -78°} \text{3-chloronia[4.3.0]bicyclononane } (Cl^+) \xrightarrow{-10°} \text{1-}CH_3\text{-2-}CH_2Cl\text{-cyclohexyl cation}$$

CHAPTER 11

Stability and Chemical Reactivity of Alicyclic Halonium Ions

11.1 STABILITY OF CYCLIC HALONIUM IONS

The stability of halonium ions as long-lived species depends on two types of stability:

1. Thermodynamic stability, as expressed, for example, by ΔH (heat of formation) from their precursors.
2. Kinetic stability, which in the absence of side reactions generally can be determined by the rate of reation with suitable nucleophiles.

The heats of formation of a series of methyl-substituted ethylene and tetramethylenebromonium ions from the appropriate dihaloalkane precursors were measured calorimetrically by Larsen and Metzner[181] in 11.5 mole % SbF_5-FSO_3H solution at $-60°$. Data are summarized in Table 40. The

TABLE 40
Relative Heats of Formation of Bromonium Ions (ΔH_{obs}) in 11.5 Mole % SbF_5 According to Larsen and Metzner[181]

Precursor	Ion	ΔH_{obs}
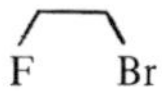 F Br	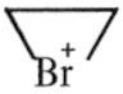 Br^+	$\sim\Delta.0$
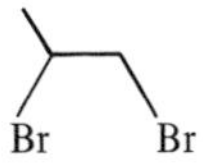 Br Br	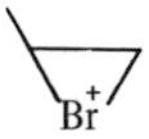 Br^+	-3.29 ± 0.51
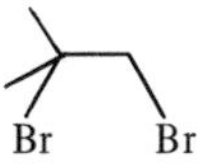 Br Br	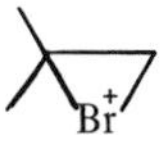 Br^+	-8.39 ± 0.87
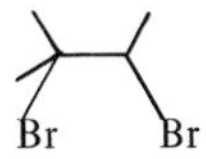 Br Br	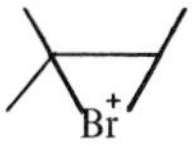 Br^+	-13.4 ± 0.90
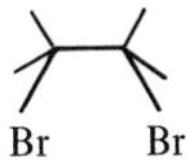 Br Br	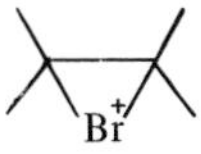 Br^+	-22.9 ± 4.00
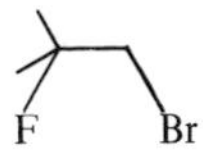 F Br	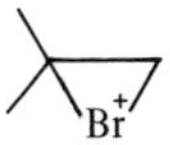 Br^+	-11.3 ± 1.83
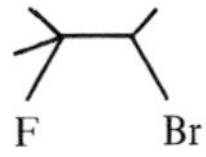 F Br	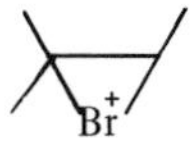 Br^+	-15.5 ± 2.15

TABLE 40
Relative Heats of Formation of Bromonium Ions (ΔH_{obs}) in 11.5 Mole % SbF_5 According to Larsen and Metzner[181] (continued)

Precursor	Ion	ΔH_{obs}
Br, Br	$\overset{+}{Br}$	−13.8 ± 1.87
Br, Br	$\overset{+}{Br}$	−13.3 ± 1.93

tetramethylenebromonium ion is about as stable as the *tert*-butyl cation, and is ca. 10 kcal/mole more stable than a correspondingly substituted ethylenebromonium ion. The tetramethylenechloronium ion is ca. 7.5 kcal/mole less stable than its bromo analog.

The stability of the three-membered-ring bromonium ions seems more sensitive to methyl substitution than that of five-membered-ring ions. With both bromine and fluorine as the leaving group, substitution of a methyl for a hydrogen atom results in about a 5-kcal/mole increase in the stability of the ion for all ions except the tetramethylethylenehalonium ions. Calculations indicate that increasing strain in the starting dihalide with increasing methyl substitution is not the source of this effect. The strain energy of 2,3-dimethyl-2,3-dibromobutene may be enough to explain the large exothermic heat of formation of the tetramethylethylenebromonium ion. This would require, however, a larger value for gauche $Br\text{-}CH_3$ interactions than that given by Benson et al.,[182] There seems to be no reason to expect exceptional stabilization of the cation by the fourth methyl group. A better knowledge of the heats of formation of haloalkanes would be helpful in interpreting this value. It is worth noting that the heat of formation of the ethylenebromonium ion was estimated by Larsen to be about 2 kcal/mole. This is quite consistent with the observed substituent effect. While errors in the heats of formation of five-membered-ring ions can be significant, it is apparent that the effect of a methyl substituent is considerably reduced.

11.2 THE ROLE OF σ- AND π COMPLEXES IN THE ADDITION OF HALOGEN TO ALKENES AND ALKYNES

Three-membered-ring alkenehalonium ion intermediates (σ complexes) have been assumed in the electrophilic addition of halogen to alkenes for many years, based on the high stereospecificity of the reactions.[4-7,164] Their existence has been confirmed more recently by their preparation and spectroscopic observation (by both σ and π routes) under stable ion conditions.

Other mechanistic paths have also been suggested, including charge transfer complexes (molecular complexes)[183] and π complexes.[184,185] In electrophilic aromatic substitution, kinetic and product distribution data (i.e., consideration of substrate and positional selectivity) have shown that the position of the transition state of highest energy is not fixed but can resemble, depending on reaction conditions and the nature of the electrophile, either the most stable intermediates (σ complexes) or the starting material (π complexes).[185] The generalized concept of electrophilic reactions was discussed by Olah[181] on the basis extensive studies of the nature of intermediate carbocations.

The formation of alkene-bromine π complexes has been suggested to explain both the kinetic[188,189] and spectroscopic[190,191] behavior of these systems. In addition, a low-temperature thermographic study of the propene-bromine system has also been reported,[192] in which repeated freeze-thaw cycles were employed to increase the concentration of the complex so that "upon warming a virtually explosive reaction" occurred. Also related is the report[188] that considerable rate enhancement occurs when bromination is carried out first by cooling an alkene-bromine system (presumably to accumulate the π complex) and then allowing it to warm up, rather than carrying out the whole reaction exclusively at the later temperature. While these observations appear to indicate the formation of intermediate bromine complexes, there was until recently no direct evidence available either on the nature of these complexes or on the role they may play in the bromination of alkenes.

11.2.1 Alkenes

The relative rates of bromination of alkenes in polar solvents, such as methanol, were extensively studied, most notably by Dubois.[193] The data show a close relationship of rate ratios to the electronic effect of alkyl groups, reflecting their effect on the intermediate alkenebromonium ions (σ complexes). Olah and Hockswender[194] determined the relative reactivities of the bromination of a series of representative alkenes in a low-polarity medium, 1,1,2-trichlorotrifluoroethane (Freon 113) solution under strictly controlled conditions in an all Teflon apparatus (the reactions are sensitive to external effects including that of glass). The reactivities compared to ethene were determined by the competitive technique at -35° and are summarized in Table 41, along with those of Dubois[193] for the bromination of the same alkenes in methanol.

The rate of reaction of alkenes with bromine in 1,1,2-trichlorotrifluoroethane solution is too fast to measure by noncompetitive methods. However, the competitive method of rate determination overcomes this difficulty, and the data obtained preclude either diffusion- or encounter rate-controlled reactions.

TABLE 41
Relative Rates of the Addition of Bromine to Alkenes Compared to Ethene

Alkene	Br_2 in $CF_2ClCFCl_2$, k_{rel} [a]	Br_2 in CH_3OH, k_{rel} [b]
Ethene	1.0	1.0
Propene	14	
1-Butene	20	95.7
1-Pentene	12	69.0
1-Hexene	10	65.7
3,3-Dimethylbutene	9	
2-Methylpropene	204	
trans-2-Butene	200	
cis-2-Butene	320	
cis-2-Hexene	460	2890
cis-2-Pentene	881	4160
cis-3-Hexene	846	6435
trans-3-Hexene	676	
cis-3-Methyl-2-pentene	1030	118,800
2-Methyl-2-butene	2300	
2,3-Dimethyl-2-butene	5680	924,000

[a] Olah and Hockswender.[194]
[b] In CH_3OH plus 0.2 *N* NaBr at 25°. Data of Dubois.[193]

While in 1,1,2-trichlorotrifluoroethane solution much lower substrate selectivities were observed than in more polar media, the rate differences are still incompatible with limiting encounter rate-controlled or diffusion-controlled conditions.

Data summarized in Table 41 indicate that bromination reaction in the low-polarity solvent system reflect a lesser degree of charge development in the transition state, which consequently is considered of the alkene-bromine π-complex type. This complex can be envisioned as arising from the initial interaction of the π system of the alkene with the electrophilic bromine, that is, the polarized bromine molecule via interaction with the back lobe of the antibonding orbital of bromine (Figure 17). This interaction would then lead to a three-center-bonded π complex such as proposed in general for the reaction of electrophiles with alkenes[184] and postulated for the reaction of borane with alkenes.[185] It is of particular interest to note that related complexes were indeed observed under stable ion conditions in the reaction of adamantylideneadamantane with bromine, as well as with other electrophiles.[163]

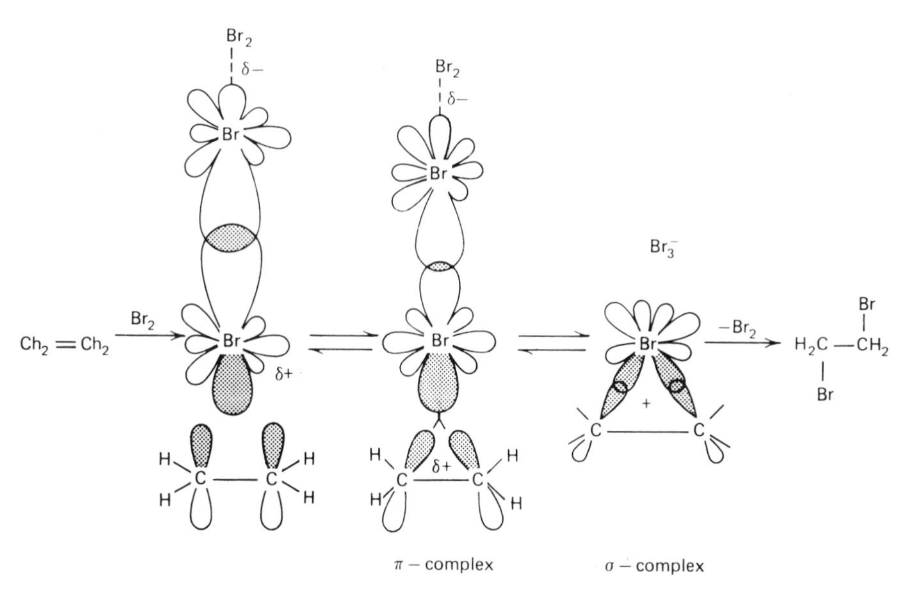

Figure 17. The path of addition of bromine to ethene.

Bromination data in the low-polarity system indicate that the transition states of the highest energy of the reactions lie early along the reaction coordinate and resemble more the starting alkenes than the intermediate alkenebromonium ions. The transition states therefore have the character of molecularly π-bonded complexes. This complex, subsequently, via cleavage of the bromine-bromine bond, and simultaneous particiaption by the non-bonded electron pairs of the rehybridized bromine atom with the developing carbenium center, forms the three-membered-ring bromonium ion intermediate (σ complex) (Figure 17). The latter alkylenebromonium ion is then displaced by a bromide ion from the accessible backside, accounting for the observed exclusive trans stereospecificity of the bromine addition reactions. This is a reaction path similar to that found in electrophilic substitution reactions of benzene and alkylbenzenes with strong electrophiles,[186,187] and is well-differentiated from reactions showing "late" transition-state character, with the transition state of highest energy resembling alkenebromonium ion intermediates (or σ complexes). If the transition state occurred early along the reaction coordinate, similar to that in electrophilic aromatic substitutions with strong electrophiles, it would be expected that there would be less demand for its stabilization by alkyl substituents, and rate differences would more closely parallel the π-donor ability of the corresponding alkenes instead of the stability of the related alkylene-bromonium ions (σ complexes).

Although it is difficult to compare directly the stabilities of π and σ complexes of alkenes with bromination rates of the same alkenes, comparisons can be made with available data on silver alkene π complexes on the one hand and the stabilities of cyclic alkylenebromonium ions on the other. Although the stability constants of Ag^+ complexes (summarized in Table 42) show some dependence on a composite of steric, electronic, and solvent effects, the importance of these values lies in the fact that they cover only a very limited range and show only a slight increase in complex stability with increasing alkyl substitution.

In contrast, as shown in Table 42, based on the data of Larsen, the heats of formation of σ-bonded, three-membered-ring ethylenebromonium ions, generated from the corresponding vicinal dibromides, show a striking effect of alkyl substitution. Although it is generally accepted that a more stable intermediate ion (a carbocation or a halonium ion) would be formed faster under kinetic conditions, thermochemical data do not allow a direct conversion of relative stability into relative rates of formation (from an alkene and bromine). How-

TABLE 42
Heats of Formation of Alkenebromonium Ions[a] and the Relative Stability of Alkene-Silver π Complexes[b]

Ion	ΔH (kcal/mole)[c]	π complex	Relative stability constant
CH_2—CH_2 bridged by Br^+	+1.0	CH_2=CH_2 · Ag^+	1.0
CH_3CH—CH_2 bridged by Br^+	-3.3	CH_3CH=CH_2 · Ag^+	0.41
$(CH_3)_2C$—CH_2 bridged by Br^+	-11.3	$(CH_3)_2C$=CH_2 · Ag^+	0.18
$(CH_3)_2C$—$CHCH_3$ bridged by Br^+	-15.5	$(CH_3)_2C$=$CHCH_3$ · Ag^+	0.36
$(CH_3)_2C$—$C(CH_3)_2$ bridged by Br^+	-22.9	$(CH_3)_2C$=$C(CH_3)_2$ · Ag^+	0.005

[a]Measured at −60° in 11.5 mole % SbF_5 in FSO_3H by Larsen.[181]
[b]Ref. 195.
[c]Heats of formation of halonium ions (and anions) are from the corresponding vicinal dihalide precursors (in 11.5 mole % SbF_5 in FSO_3 at −60°). Since the heat of formation of the anion produced is not known, ΔH is a relative quantity. Differences in ΔH represent differences in the heat of formation of the ions from the dihalide precursors in solution with the same leaving groups (anions).

ever, data of Table 41 clearly indicate that a very wide range in order of magnitude of relative rates would be expected for a reaction in which the transition state occurs late along the reaction coordinate and thereby resembles in nature the cyclic bromonium ion. This is more the case in the work by Dubois[193].

While neither the measured bromonium ion nor silver ion complex stabilities are directly comparable with rates of bromination of alkenes,[196] they indicate that the relative rates observed in low polarity systems are much more in accord with a π-complex-type transition state for the reaction than the σ-complex mechanism proposed in polar solvent systems. It has been shown for electrophilic aromatic substitutions that a similar change in relative rates is observed in reactions involving either early (π-complex-type) or late (σ-complex-type) transition states, illustrating how relative-rate data reflect the nature of the transition states in the reaction.

While previously it was assumed that a π-complex-type mechanism for alkene additions would result only in negligible spread of relative-rate values (corresponding to silver ion-alkene π-complex stability constants[183] or iodine-alkene stability data,[197] since both are qualitatively similar), the differences observed in brominations in low-polarity solvents[189,194] do not support this assumption. Theoretical calculations[198] are also in accord with this suggestion and indicate that significant differences are to be expected. These calculations on systems involving an alkene and an interacting electrophile show that only a week interaction exists between ethene and silver ion or molecular chlorine, but that significant stabilization in the latter case can be obtained by increased chlorine-chlorine bond length and partial charge development. Therefore it seems likely that since iodine, like silver ion, forms only a weak complex with alkenes, a strict correspondence between measured stability constants and relative rates is not expected. However, because the halogen-halogen bond is not greatly deformed and the alkene and halogen are in close proximity, some approximate correlation seems reasonable.

In addition, the development of a partial charge would allow a general parallelism between increased alkyl substitution and increased reactivity, as observed in the bromination and in other electrophilic reactions of alkenes.

The relative reactivities of the bromination of cyclohexene, cyclopentene, and norbornene were also determined. The results are listed in Table 43 along with, for comparison, some other reported data on additions to the same cycloalkenes. In the case of the more strained alkenes, higher rates are expected with increasing ring size, which should result in greater relief of strain in the transition state, relative to a smaller size ring.[199] The data in Table 43 do seem to support this hypothesis, and relative-rate data of the studied brominations indicate that the transition states of highest energy of the reactions are early and resemble the starting alkenes.[200]

TABLE 43
Relative Rates of Addition to Cycloalkenes[a]

Reagent	Cyclopentene	Norbornene
Bromine	4.2	36
Chromyl chloride	4.9	312
Ag^+	2.0	17
Diethyl aluminum hydride	9.4	
Ozone	3.9	4.3
Osmium tetroxide	22	72
Diimide	15.5	450

[a]Cyclohexene = 1.0.

11.2.2 Alkynes

The relative rates of bromine addition to some alkyl-substituted alkynes in 1,1,2-trichlorotrifluoroethane solution at 0° were determined.[194] These results are listed in Table 44, along with related data obtained by Pincock and Yates[201] for bromination in acetic acid. Temperature effects on the bromination in acetic acid lead to a large negative entropy effect, which was interpreted as being due to the involvement of an intermediate alkynebromonium ion. Although the range of rates reported by Pincock and Yates is not very broad, bromination in 1,1,2-trichlorotrifluoroethane does lead to reduced selectivity, as also noted for alkenes. In addition, while alkynes react at rates similar to, or even greater than, those of the corresponding alkenes in some electrophilic reactions, for reactions that involve strained bridged (cyclic) transition states, they react up to 10^5 times more slowly then alkenes.[202] Similar behavior was also noted in a bromination study in which alkynes were found to react more slowly by at least a factor of 10^4 than the corresponding alkenes. Although no corresponding π- and σ-complex stability data are available for alkynes, the data summarized in Table 44 are also considered to indicate in the bromination of alkynes a mechanism with an early transition state of a π-complex nature, similar to that of the related bromination of alkenes.

The difference in behavior of alkenes and alkynes is quite reasonable when one considers that the decreased reactivity of alkynes results from an increase in the energy required for initial π-complex formation. Since the unsaturated bond in alkynes has cylindrical symmetry along the carbon-carbon bond axis and involves a measure of electron delocalization,[203] total electron availability for complex formation is reduced. This then requires increased energy for π-complex formation and results in slower rates of reaction.

TABLE 44
Relative Rates of Addition of Bromine to Alkynes

Alkyne	Br_2 in CF_2ClCCl_2F[a]	Br_2 in CH_3COOH[b]
3,3-Dimethylbutyne	1.0	1.0
2-Butyne	1.6	
1-Hexyne	2.4	0.61
2-Hexyne	3.0	
3-Hexyne	3.2	33.6

[a]For the reaction at 0° ref 194.
[b]For the reaction at 24.8°, relative second-order rate constants.

11.3 REACTIONS OF ETHYLENEHALONIUM IONS

The chemical reactivity of ethylenehalonium ions has not yet been studied in any detail.[204] With halide ions they give vicinal dihalides with exclusive trans stereospecificity, in accordance with the long-suggested addition mechanism of halogens to olefins.

$$\text{CH}_2\overset{X^+}{—}\text{CH}_2 + Y^- \rightarrow \text{CH}_2(X)\text{–CH}_2(Y)$$

X = I, Br, Cl
Y^- = I, Br, Cl, F

Alcohols and other nucleophiles give β-haloethyl ethers and related products, in addition to elimination products.

When reacted with aromatic hydrocarbons, ethylenehalonium ions give β-haloethylarenes.

$$C_6H_6 + \text{CH}_2\overset{X^+}{—}\text{CH}_2 \rightarrow C_6H_5\text{–CH}_2\text{CH}_2\text{X}$$

The β-haloethylation reaction seems to be of general utility. Ethylenehalonium ions may also be responsible for β-haloethylations observed with 1-fluoro-2-chloro(or bromo)ethane[205]

$$C_6H_6 + \text{F-CH}_2\text{CH}_2\text{Cl (Br)} \xrightarrow{BBr_3} C_6H_5\text{–CH}_2\text{CH}_2\text{Cl (Br)}$$

Ethylene dihalides are also potential carcinogens, and their biological alkylating ability could also indicate β-halogen participation forming ethylenechloronium ion-like reactive intermediates.[204]

A more detailed study of the chemical reactivity of ethylenehalonium ions clearly is indicated and is expected to provide much interesting data.

11.4 REACTIONS OF TETRAMETHYLENEHALONIUM IONS

The synthetic utility of tetramethylenehalonium ions has been demonstrated by Peterson and coworkers.[176] The tetramethylenechloronium ion was used to alkylate methanol, acetonitrile, dimethyl ether, methyl propyl ether, acetone, and acetic acid. The initially formed intermediate open-chain ions of these

CH_3OH → $CH_3O(CH_2)_4Cl$ 70-80%

CH_3CN → $CH_3C{\equiv}\overset{+}{N}(CH_2)_4Cl$ $\xrightarrow{H_2O}$ $CH_3C(=O)NH(CH_2)_4Cl$ 86%

$(CH_3)_2O$ → $(CH_3)_2\overset{+}{O}(CH_2)_4Cl$ → $CH_3O(CH_2)_4Cl$ 91%

$CH_3OCH_2CH_2CH_3$ → $CH_3CH_2CH_2\overset{+}{O}(CH_3)(CH_2)_4Cl$ → $CH_3CH_2CH_2O(CH_2)_4Cl$ 63% + $CH_3O(CH_2)_4Cl$ 28%

CH_3COCH_3 → $CH_3C(CH_3)=\overset{+}{O}(CH_2)_4Cl$ $\xrightarrow[Na_2CO_3]{CH_3OH}$ $HO(CH_2)_4Cl$ 50%

CH_3COOH → $CH_3COO(CH_2)_4Cl$ 62%

reactions were directly observed by nmr at $-60°$. The final reaction products were obtained by further reactions (quenching), generally in fair to good yield.

However, the cleavage of unsymmetric tetramethylenehalonium ions in reactions carried out with various nucleophiles (such as methanol, dimethyl ether, acetic acid, and trifluoroacetic acid) generally gave two products. In the methanolysis of 2-halomethyltetramethylenehalonium ions major and minor products were obtained.

CH_3OH / Na_2CO_2

minor major

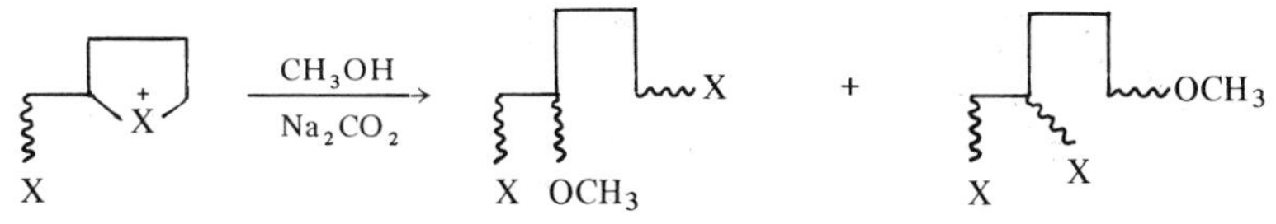

X = Cl
X = Br

Peterson and Bonazza[177] also reported the reaction of three-substituted tetramethylenechloronium ions with a variety of nucleophiles under similar conditions, according to the following equation.

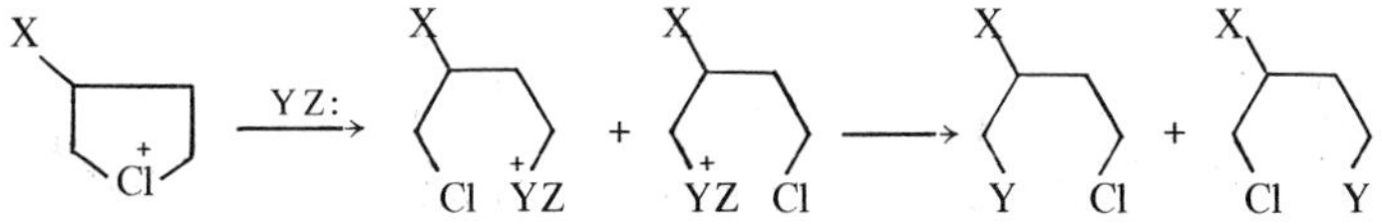

YZ:

X = CH_3
X = Cl

Y = CH_3O; Z = H, CH_3
Y = CH_3COO, CF_3COO; Z = H

CHAPTER 12

Heteroaromatic Halophenium Ions

In certain cases the heteroatom in hetarenes can also be ionic halogen. The systems studied so far centered on halophenium ions (five-membered-ring halonium analogs of thiophene[19]).

12.1 HALOPHENIUM IONS

Silver-assisted solvolysis of PhCCl=CHCH″=CBrPh in HOAc and in Ac_2O gave PhCCl=CHCH=CPh–OAc and PhCCl″CHC″CPh. Analyses of the deuterium distribution in the products from deuterated starting material showed that 24% (in HOAc) and 30% (in Ac_2O) of the reaction proceeds through a chlorophenium ion intermediate.[206] Similar results were obtained from a study of the solvolysis products.

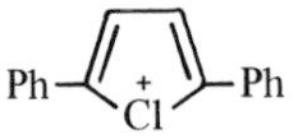

Ph Cl+ Ph

Beringer et al.[207] found that 1,4-dilithio-1,2,3,4-tetraphenyl-1,3-butadiene reacts with *trans*-chlorovinyliodoso dichloride at −78° to give, in low yield, 2,3,4,5-tetraphenyliodophenium chloride which is converted to tetraphenyliodophenium iodide by treatment with LiI.

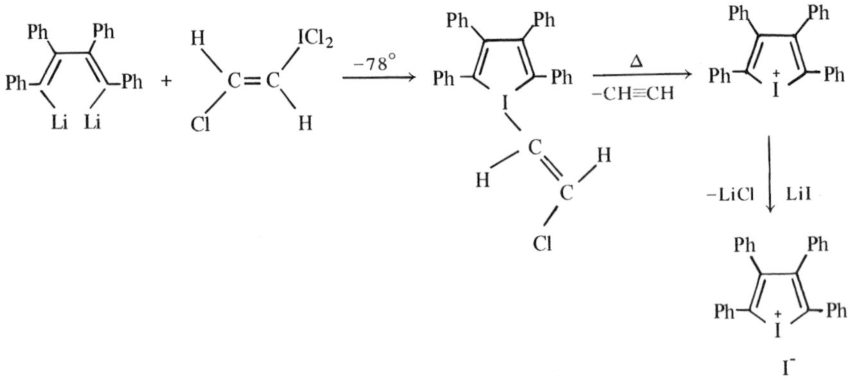

Ph Ph Ph Ph Li Li + H ICl_2 C=C Cl H —(−78°)→ Ph Ph Ph Ph I C H C H Cl —(Δ, −CH≡CH)→ Ph Ph Ph Ph I+ —(LiI, −LiCl)→ Ph Ph Ph Ph I+ I−

12.2 BENZIODOPHENIUM IONS

Beringer and coworkers[207] reported that the low-temperature reaction of *trans*-1-lithio-2-*o*-lithiophenyl-1-phenyl-1-hexene with *trans*-chlorovinyliodoso dichloride in ether gave a 26% yield of 3-butyl-2-phenylbenziodophenium chloride which was converted by metathesis with ICl to 3-butyl-2-phenylbenziodophenium iodide. It was suggested that the reaction probably proceeds via the trivalent organoiodine intermediate.

n-C_4H_9 Ph C=C Li Li + H ICl_2 C=C Cl H ⟶ n-C_4H_9 Ph I H H Cl

−HC≡CH

n-C_4H_9 Ph I^+ I^- $\xleftarrow{KI}$ n-C_4H_9 Ph I^+ Cl^-

Another five-membered halonium ion, the dibenziodophenium ion, was also prepared by Beringer and Nathan.[208] The X-ray crystal structures of the ions were also determined.

Li Li $\xrightarrow[-78^\circ]{\text{H(Cl)C=C(H)ICl}_2}$ I^+ Cl^-

+ HC≡CH

Beringer and coworkers[208] reported reactions of benziodophenium salts. Thermal decomposition of the salts at their melting points gave a mixture of *cis*- and *trans*-stilbenes.

n-C_4H_9 Ph I^+ X^- $\xrightarrow{\Delta}$

X = Cl
X = I

n-C_4H_9 Ph C=C I X
X = Cl; 50%
X = I; 58%

n-C_4H_9 X C=C I Ph
X = Cl; 50%
X = I; 42%

Reaction with sodium methoxide in anhydrous methanol gave an equimolar mixture of *cis*- and *trans*-2-iodophenyl-1-methoxy-1-phenyl-1-hexene. The mixture then underwent acid-catalyzed hydrolysis.

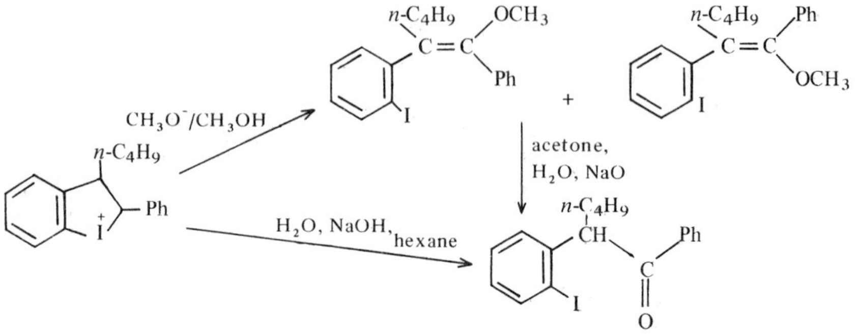

n-C_4H_9 OCH_3 Ph I + n-C_4H_9 Ph OCH_3 I

CH_3O^-/CH_3OH

n-C_4H_9 Ph $\overset{+}{I}$

acetone, H_2O, NaO

H_2O, NaOH, hexane

n-C_4H_9 CH Ph C O I

The mechanism of the reaction, which shows attack by nucleophiles exclusively at the 2-position to give almost equal amounts of *cis*- and *trans*-stilbenes, was suggested to be

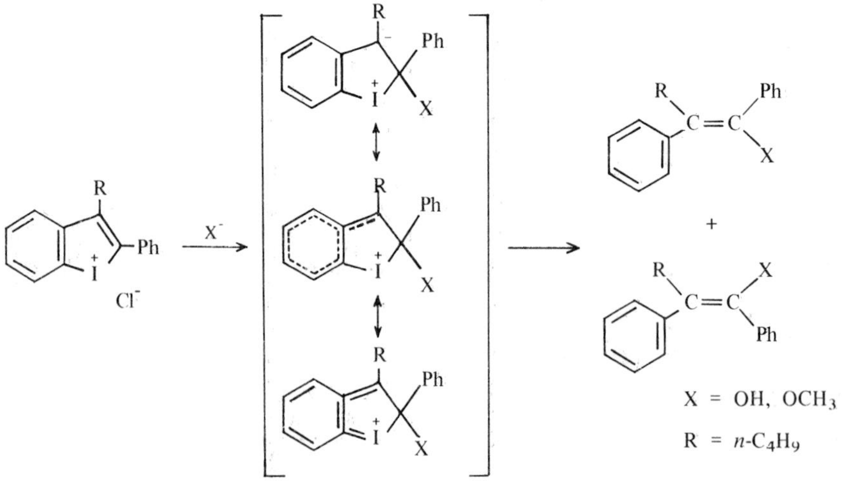

R Ph $\overset{+}{I}$ Cl^- $\xrightarrow{X^-}$ [R Ph X $\overset{+}{I}$ ↔ R Ph X $\overset{+}{I}$ ↔ R Ph X $\overset{+}{I}$] ⟶ R Ph X + R X Ph

X = OH, OCH_3

R = n-C_4H_9

Attack of the nucleophile at the 2-position from above or below the plane of the ring can occur with with equal probability, giving a dipolar intermediate. Delocalization of the negative charge into the aromatic ring and into the *d* or *f* orbitals of iodine might help to stabilize the intermediate. E_1 elimination of the I^+ group would give equal amounts of the cis and trans isomers.

When benziodophenium ions are treated with water in the presence of oxygen, oxidative hydrolysis occurs rapidly (ca. 1-2 hr). Under the same conditions but in the absence of oxygen, a very slow (ca. 1-month) hydrolysis occurs in almost quantitative yield. Reduction with either sodium borohydride in ethanol at room temperature or sodium iodide in acetone gives the hydroxy ketone. The hydroxy ketone is oxidatively cleaved with lead tetraacetate to benzoic acid and ketone, which are also formed by thermal cleavage of the intermediate oxidation product.

$R = n\text{-}C_4H_9$

$H_2O\text{-}O_2$, fast; + HCl; $NaBH_4$-ethanol or NaI-acetone; Δ; H_2O, no O_2, slow, −HCl; PhCOOH +

12.3 BENZCHLOROPHENIUM IONS

In an attempt to prepare and study benzchlorophenium ions, *o*-(β,β-dichloroethenyl)phenyl- and *o*-(α-methyl-β,β-dichloroethenyl)phenyldiazonium fluorophosphates were used as precursors by Olah and Yamada.[19] In their thermal decomposition the desired benzchlorophenium ion could not be isolated, but *o*-(β-chloroethynyl)chlorobenzene was obtained in good yield. Its formation

Δ, benzene; $PF_6^{\ominus}$

clearly shows that the benzchlorophenium ion is an intermediate which via ring opening and deprotonation gives the observed product. The presence of the highly acidic proton at C-3 facilitates ring opening of the benzchlorophenium ion (and subsequent deprotonation).

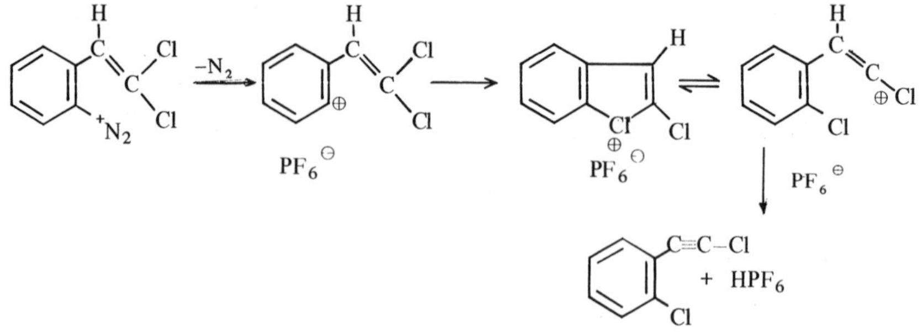

If the acidic proton is absent in the benzchlorophenium ion, a different behavior is observed. The thermal decomposition of *o*-(α-methyl-β,β-dichloroethenyl)phenyldiazonium fluorophosphate in refluxing benzene (for 0.5 hr) gave a mixture of 1-chloro-1-fluoro-2-(*o*-chlorophenyl)pentene-1 and 1,1-dichloro-2-(*o*-fluorophenyl)propene-1 (25.6%), 1-chloro-1-(*o*-chlorophenyl)-2-fluoropropene-1 (24.5%), and 1,1-dichloro-2-(2-biphenyl)propane-1 (46.5%).

The formation of 1-chloro-1-fluoro-2-(*o*-chlorophenyl)propene-1 and 1-chloro-1-(*o*-chlorophenyl)2-fluoropropene-1 shows the presence of the 2-chloro-3-methylbenzochlorophenium ion as the intermediate, since only its ring opening can introduce chlorine into the ortho position of the phenyl ring. Additional products observed arise from the incipient phenyl cation.

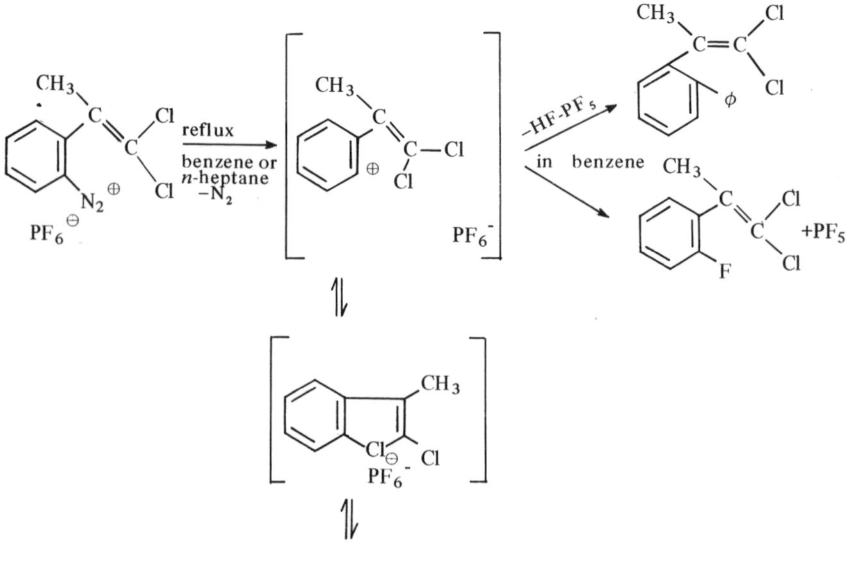

CH3 C C–F Cl Cl trans ← in n-heptane –PF5 [CH3 C C–Cl ⊕ Cl PF6⁻] → in benzene –PF5 CH3 C=C Cl F cis Cl + CH3 C=C F Cl Cl trans

↓ aryl migration

[CH3–C⊕ C–Cl Cl PF6⁻] → in benzene or n-heptane –PF5 CH3 C F C~Cl Cl

12.4 DIBENZHALOPHENIUM IONS

The dibenziodophenium ion containing an aromatic five-membered iodonium ring was first prepared from 2-iodosobiphenyl[209]

IO + HX → ⊕ I X⊖

2,2′-Diiodoso-, and 2,2′-diiodosodichlorobiphenyl were also treated with water and SO_2 to give dibenziodophenium ions.[210] However, 2,2′-diiodobiphenyl,

IO IO or ICl_2 ICl_2 → ⊕ I X⊖

when treated with peracetic acid (to give the corresponding diiodosoacetate derivative) and subsequently with H_2SO_4 and NaCl, gave not only the dibenziodophenium ion but also a bisiodonium ion.[211] The reactions are indicated to proceed via three-valent iodine compounds that is, R–ICl_2, R–IO, $RI(OAc)_2$. However, the dibenziodolphenium ion was also prepared from diazotized 2,2′.

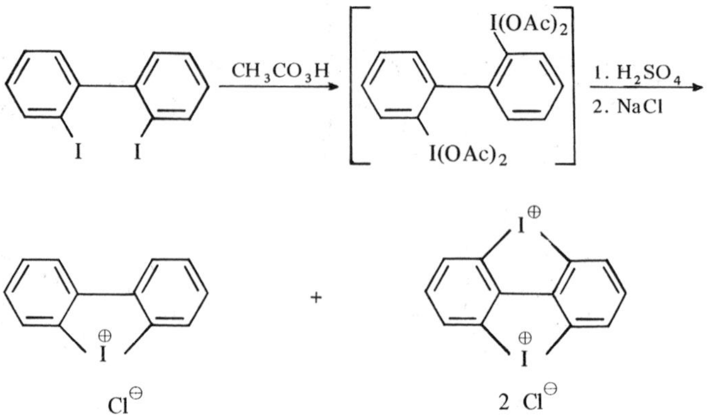

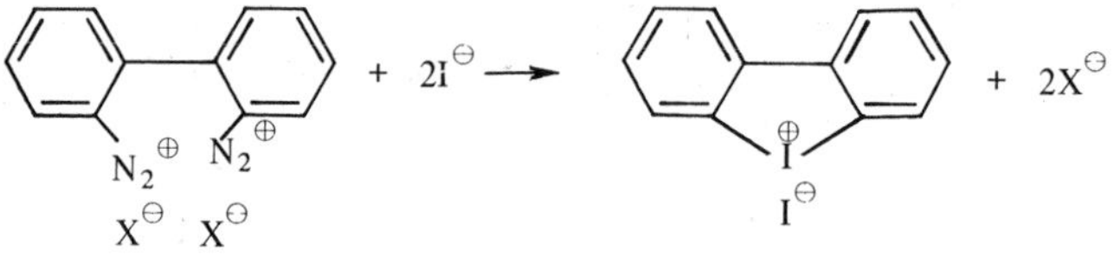

diaminobiphenyl and iodide ion.[212] This reaction could proceed in the following manner.

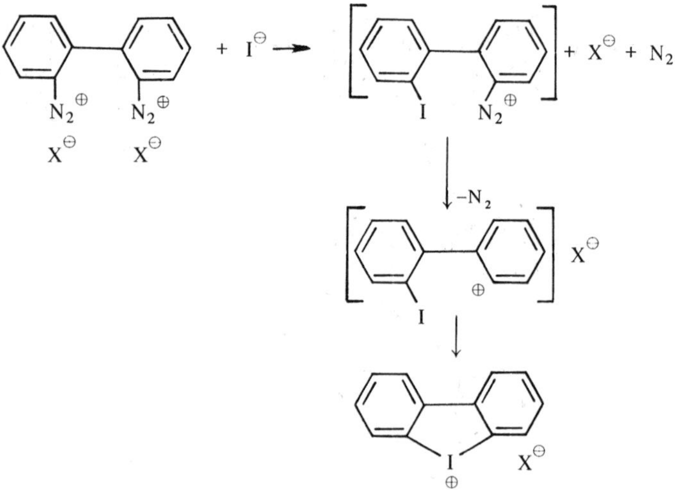

Heaney[213] and Sandin[214] prepared the dibenziodophenium ion by thermolysis of 2-(2′-iodobiphenyl)diazonium fluoroborate. The dibenziodophenium ion react

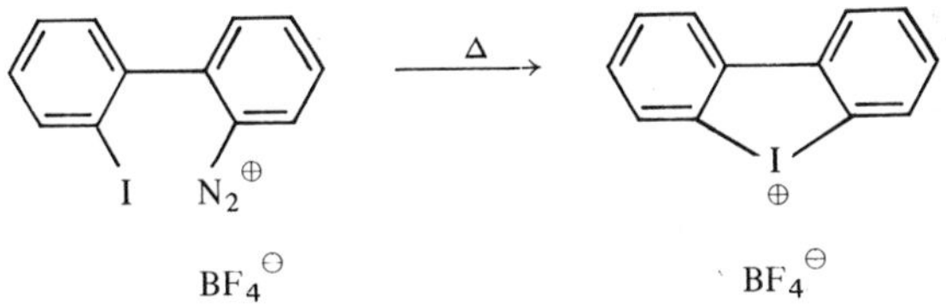

with phenyllithium to give 5-phenyl-5*H*-dibenziodole, which is stable at room temperature for several days.[215] The phenyl-iodine bond is easily cleaved by

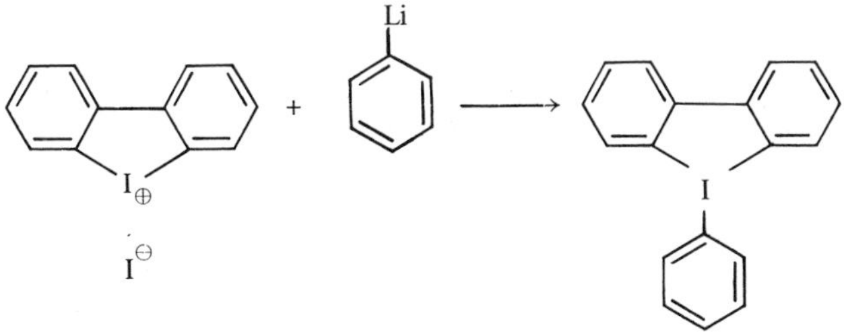

hydrogen chloride, iodine, and triphenylboron to give the dibenziodophenium ion.

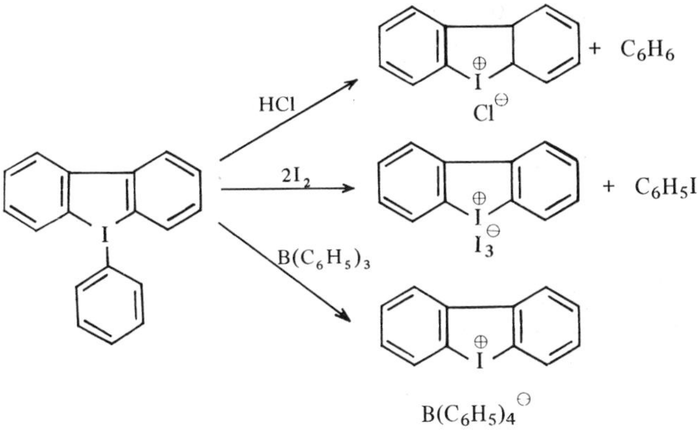

The thermal reaction of 5-phenyl-5*H*-dibenziodole in refluxing hexane gives 2-diodo-2′-phenylbiphenyl (formed via an ionic mechanism).

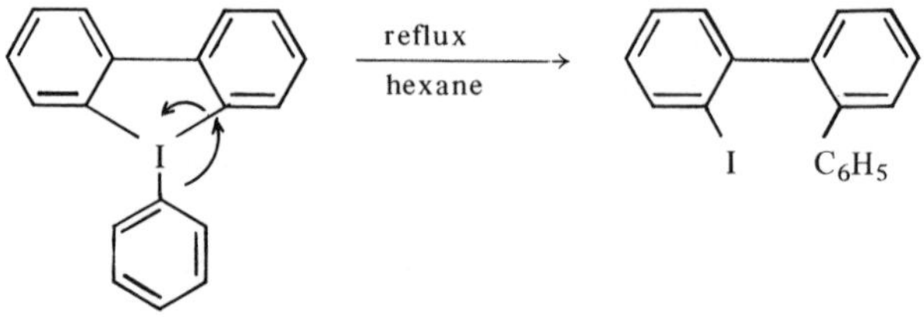

reflux
hexane
I C_6H_5

In the reaction in hexane at room temperature, however, not only the ionic but also a radical reaction occurs,[216,217] giving the following products.

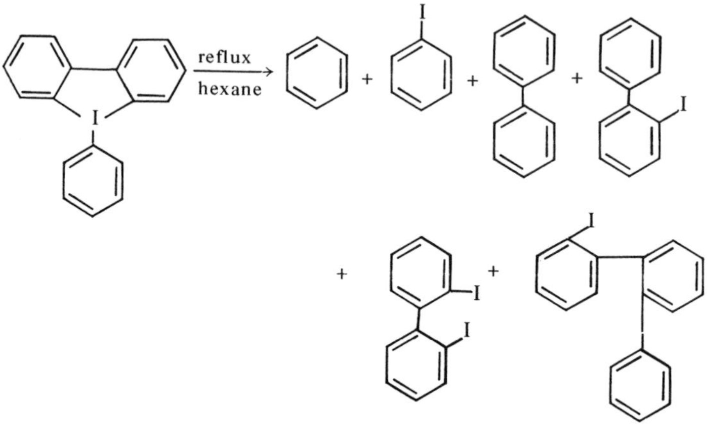

reflux
hexane

The reaction proceeds in the following way.

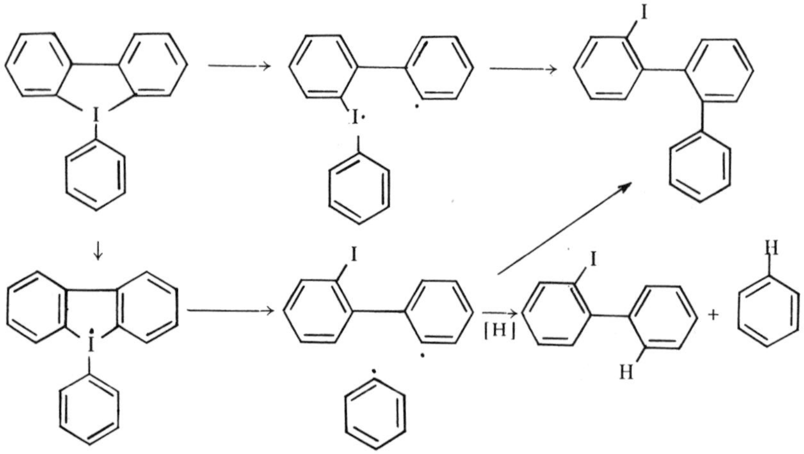

[H]

In the presence of protic or Lewis acids the reaction proceeds by the heterolytic mechanism to give cyclic and acyclic iodonium ions in good yield.[217]

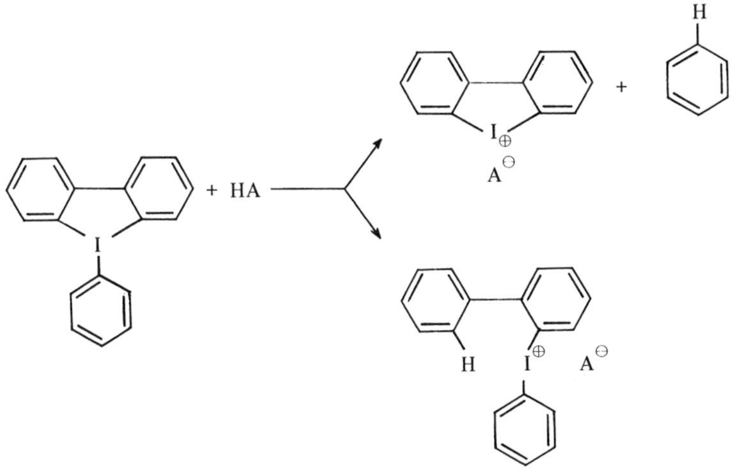

Dibenziodophenium sulfate reacts with acetate ion to give 2-iodo-2′-biphenyl acetate. In the reaction with methoxide ion, the products are not 2-iodo-2′-methoxybiphenyl but biphenyl, 2-iodobiphenyl, and 2,2′-diodobiphenyl. This suggests that the reaction did not proceed heterolytically but homolytically.[218] Usually, the reactivity of the dibenziodophenium ion is lower than that of diphenyliodonium ions, which can be explained by the aromatic five-membered iodonium ring (the "iodophene" system).

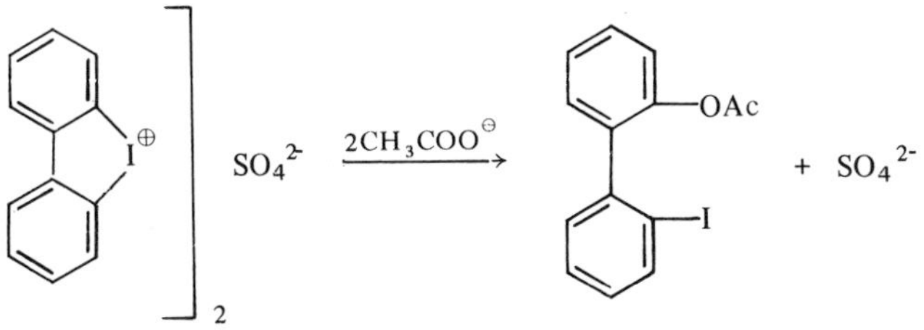

Six-, seven- and eight-membered-ring iodonium ions have also been synthesized, as were six-membered-ring iodonium ions containing hetero atoms.[11]

n = 1, 2, 3

X = O, S, NH

Sandin and Hay[214] synthesized five-membered-ring cyclic bromonium and chloronium ions from the corresponding 2-amino-2′-halobiphenyls in aqueous solution.

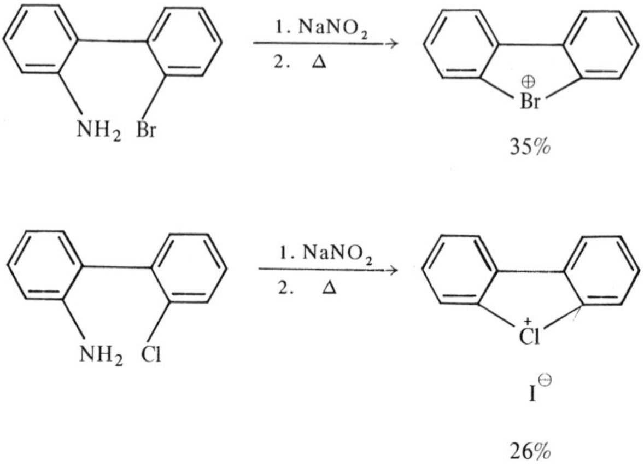

NH_2 Br

1. $NaNO_2$
2. Δ

$Br^{\oplus}$

35%

NH_2 Cl

1. $NaNO_2$
2. Δ

Cl^{+}

$I^{\ominus}$

26%

Heany et al.[213] prepared the same halonium ions from the corresponding diazonium salts in benzene solution and isolated them in good yield.

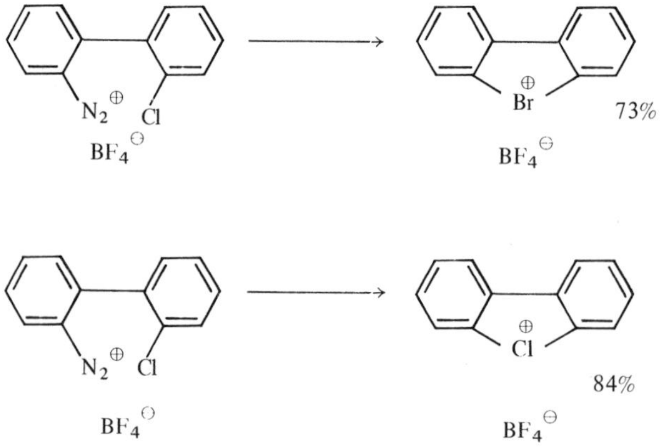

$N_2^{\oplus}$ Cl

$BF_4^{\ominus}$

$Br^{\oplus}$

$BF_4^{\ominus}$

73%

$N_2^{\oplus}$ Cl

$BF_4^{\ominus}$

$Cl^{\oplus}$

$BF_4^{\ominus}$

84%

CHAPTER 13

Conclusions and Future Outlook

Halonium ions are not only the reaction intermediates in electrophilic halogenations, but also represent in their own right a significant class of organic onium compounds of substantial preparative and chemical structural interest. Nearly all types of halonium ions with structures related to other analogous onium ions (acyclic as well as cyclic or even bicyclic) can now be prepared. Halonium ions can be prepared *in situ* or isolated as stable salts for further reactions. Recent developments mostly emphasized the preparation and spectroscopic study of halonium ions. Their utility as reagents in organic syntehsis is in its formative stage and can be expected to undergo fast development.

Alkyl halides are some of the most useful laboratory alkylating agents. Aryl halides, however, are considerably less reactive. However, diphenylhalonium ions (especially iodonium ions) are reactive phenylating agents. That is, the bond between the phenyl ring and the halonium center is easily cleaved on reaction with a nucleophile to give phenylated products. Iodonium ions bearing a thiophene, furan or pyridine ring as a ligand are expected to be able to transarylate nucleophiles via transferring the hetercyclic group. Iodonium ions bearing olefinic or acetylenic groups as ligands are also expected to have significant reactivity. Iodonium ylides ($-I^{\oplus}-\overset{\ominus}{C}<$ or $-I^{\oplus}-\overset{\ominus}{N}-$) are interesting reactive intermediates for carbenes or nitrenes.

Organic halonium ions and their related ylides, although still in their infancy as compared with such well-developed classes of onium ions as ammonium, oxonium, sulfonium, and phosphonium ions, nevertheless, promise substantial utility and significance. It is hoped that this review will further interest in and future development of the fascinating field of organic halonium ions.

References

1. C. Hartmann and V. Meyer, *Chem. Ber.*, **27**, 426 (1894).

2.a. A. N. Nesmejanov, L. G. Makarova, and T. P. Tolstoya, *Tetrahedron*, **1**, 145 (1957).

2.b. A. N. Nesmeyanov, *Selected Works in Organic Chemistry*, Pergamon Press, Oxford, 1963, and references therein.

3. G. A. Olah and J. R. DeMember, *J. Am. Chem. Soc.*, **91**, 2113 (1969).

4. I. Roberts and G. E. Kimball, *J. Am. Chem. Soc.*, **59**, 947, (1937).

5.a. B. Capon, *Quart. Rev.* (London), **18**, 45 (1964).

5.b. R. C. Fay, in *Topics in Stereochemistry*, E. L. Eliel and N. L. Allinger, eds., Wiley-Interscience, New York, Vol. 3, 1968, p. 237.

6. S. Winstein and H. J. Lucas, *J. Am. Chem. Soc.*, **61**, 1576, 2845 (1939).
7. J. C. Traynham, *J. Chem. Ed.*, **40**, 392 (1963).
8. E. Gould, *Mechanism and Structure in Organic Chemistry*, Holt. Rinehart, Winston, New York, 1959, p. 523.
9. G. A. Olah and J. M. Bollinger, *J. Am. Chem. Soc.*, **89**, 4744 (1967).
10. P. E. Peterson, *Acc. Chem. Res.*, **4**, 407 (1971), and references therein.
11.a. For reviews of earlier work, D. F. Banks, *Chem. Rev.*, **66**, 243 (1966).
11.b. R. B. Sandin, *Chem. Rev.*, **32**, 249 (1943).
12. G. A. Olah and E. G. Melby, *J. Am. Chem. Soc.*, **94**, 6220 (1972).
13. G. A. Olah, Y. K. Mo, E. G. Melby, and H. C. Lin, *J. Org. Chem.*, **38**, 367 (1973).
14.a. G. A. Olah and J. M. Bollinger, *J. Am. Chem. Soc.*, **90**, 947 (1968).
14.b. G. A. Olah, J. M. Bollinger, and J. M. Brinich, *J. Am. Chem. Soc.*, **90**, 2587 (1968).
15. G. A. Olah and P. E. Peterson, *J. Am. Chem. Soc.*, **90**, 4675 (1968).
16. P. E. Peterson, B. R. Bonazza, and P. M. Hendrichs, *J. Am. Chem. Soc.*, **95**, 2222 (1973).
17. P. E. Peterson and W. F. Boron, *J. Am. Chem. Soc.*, **93**, 4076 (1971).
18. J. H. Exner, L. D. Kershner, and T. E. Evans, *Chem. Commun.*, 361 (1973).
19. G. A. Olah and Y. Yamada, *J. Org. Chem.*, **40**, 1107 (1975), and references therein.
20. IUPAC, *Definitive Rules for Nomenclature of Organic Chemistry*, Sections A and B, Butterworths, London, 1966, Rule 4.
21. IUPAC, *Definitive Rules for Nomenclature of Organic Chemistry*, Butterworths, London, 1966, Rules B-4 and C82.3.
22. P. E. Peterson and B. R. Bonazza, *J. Am. Chem. Soc.*, **94**, 5017 (1972).
23. G. A. Olah, G. Liang, and J. Staral, *J. Am. Chem. Soc.*, **96**, 8112 (1974).
24. G. A. Olah, Y. Yamada, and R. J. Spear, *J. Am. Chem. Soc.*, **97**, 680 (1975).
25. G. A. Olah and T. E. Kiovsky, *J. Am. Chem. Soc.*, **89**, 5692 (1967).
26.a. G. A. Olah and J. R. DeMember, *J. Am. Chem. Soc.*, **92**, 718, (1970).
26.b. G. A. Olah and J. R. DeMember, *J. Am. Chem. Soc.*, **92**, 2562 (1970).
26.c. G. A. Olah, J. R. DeMember, Y. K. Mo, J. J. Svoboda, P. Schilling, and J. A. Olah, *J. Am. Chem. Soc.*, **96**, 884 (1974).
26.d. G. A. Olah, P. W. Westerman, E. G. Melby and Y. K. Mo, *J. Am. Chem. Soc.*, **96**, 3565 (1974).
27. G. A. Olah and Y. K. Mo, *J. Am. Chem. Soc.*, **96**, 3560 (1974).
28. G. A. Olah, Y. K. Mo, E. G. Melby, and H. C. Lin, *J. Org. Chem.*, **38**, 367 (1973).
29.a. G. A. Olah and Y. K. Mo, *J. Chem. Soc.*, **94**, 6864 (1972).
29.b. G. A. Olah, R. Renner, P. Schilling, and Y. K. Mo, *J. Am. Chem. Soc.*, **95**, 7686 (1973).
30.a. D. M. Grant and E. G. Paul, *J. Am. Chem. Soc.*, **86**, 2984 (1964).
30.b. D. K. Dalling and D. M. Grant, *J. Am. Chem. Soc.*, **89**, 6612 (1967).
31. J. B. Stothers, *Carbon-13 NMR Spectroscopy*, Academic Press, New York, New York, 1972; a. Table 3.1; b. Table 5.4; c. Table 5.5; d. Table 5.58; e. p. 201; f. p. 202; g. p. 55.
32. G. A. Olah and P. W. Westerman, *J. Org. Chem.*, **38**, 1986 (1973).
33.a. R. Vogel-Högler, *Acta Phys. Austriaca*, **1**, 311 (1947).
33.b. H. Siebert, *Z. Anorg. Allg. Chem.*, **271**, 65 (1953).
33.c. G. Herzberg, *Molecular Spectra and Molecular Structure*, Vol. II, D. Van Nostrand, Princeton, N. J., 1945, P. 315.
33.d. D. R. Allkin and P. J. Hendra, *Spectrochim. Acta*, **22**, 2075 (1966).
34. H. Meerwein, in D. Müller, Ed., *Houben-Weyl's Methoden der Organischen Chemie*, 4th ed., Vol. 6, No. 3, Georg Thieme, Stuttgart, 1965, pp. 329-363.

35. G. A. Olah, *J. Am. Chem. Soc.*, **94**, 808 (1972).
36. G. A. Olah, J. R. DeMember, R. H. Schlosberg, and Y. Halpern, *J. Am. Chem. Soc.*, **94**, 156 (1972).
37. G. A. Olah, A. M. White, J. R. DeMember, A. Commeyras, *J. Am. Chem. Soc.*, **92**, 4627 (1970), and references therein.
38. P. v. R. Schleyer, R. C. Fort, Jr., W. E. Watts, M. B. Comisarow, and G. A. Olah, *J. Am. Chem. Soc.*, **86**, 4195 (1964).
39. G. A. Olah and G. Liang, *J. Am. Chem. Soc.*, **97**, in press.
40. R. D. Wieting, R. H. Staley, and J. L. Beauchamp, *J. Am. Chem. Soc.*, **96**, 7554 (1974).
41. G. A. Olah, *Makromol. Chem.*, **175**, 1039 (1974).
42. P. H. Plesch, IUPAC Symposium on Macromolecules, Helsinki, July 1972.
43. J. P. Kennedy, ACS Symposium, Cleveland, Ohio, May 1973.
44. F. DeHaan, H. C. Brown, D. C. Conway, and M. G. Gibley, *J. Am. Chem. Soc.*, **91**, 4854 (1969).
45. D. G. Walker, *J. Phys. Chem.*, **64**, 939 (1960).
46. H. C. Brown and W. J. Wallace, *J. Am. Chem. Soc.*, **75**, 6279 (1953).
47. H. H. Perkampus and E. Baumgarten, *Ber. Bunsenges. Phys. Chem.*, **68**, (1964).
48. E. Kinsella and J. Coward, *Spectrochim Acta, Part A*, **24**, 2139 (1968).
49. B. Rice and K. C. Bond, *Spectrochim Acta*, **20**, 721 (1964).
50. W. v. E. Doering, M. Saunders, H. G. Boyton, H. W. Earhart, E. F. Wadley, W. E. Edwards, and G. Laber, *Tetrahedron*, **4**, 178 (1958).
51. T. B. Dence and J. D. Roberts, *J. Org. Chem.*, **33**, 1251 (1968).
52. V. V. Lyalin, V. V. Orda, L. A. Alekseeva, and L. M. Yagupoliskii, *Z. Org. Khim.*, 7, 1473 (1971); *Chem. Abstr.*, **75**, 140390a (1971).
53. G. A. Olah and E. G. Melby, *J. Am. Chem. Soc.*, **94**, 6220 (1972).
54. F. m. Beringer, M. Drexler, E. M. Gindler, and D. C. Lumpkin, *J. Am. Chem., Soc.*, **75**, 2705 (1953); F. M. Beringer, R. A. Falk, M. Karniol, I. Lillien, G. Masulio, M. Mausner, and E. Summer, *ibid.*, **81**, 342 (1959).
55. W. K. Hwang, *Sci. Sinica* (Peking), **6**, 123 (1957); *Chem. Abstr.*, **51**, 16476 (1957.).
56. I. Masson and E. Race, *J. Chem. Soc.*, 1718, (1937).
57. F. M. Beringer, H. E. Bachofner, R. A. Falk, and M. Leff., *J. Am. Chem. Soc.*, **80**, 4279 (1958); F. M. Beringer and R. A. Falk, *Ibid*, **81**, 2997 (1959).
58. F. M. Beringer and I. Lillien, *J. Am. Chem. Soc.*, **82**, 725 (1960).
59. F. M. Beringer, J. W. Dehn, Jr., and M. Winicov, *J. Am. Chem. Soc.*, **82**, 2948 (1960).
60. A. M. Nesmeyanov, T. P. Tolstaya, and I. M. Lisichkina, *Izv. Akad. Nauk SSSR, Ser. Khim.*, 194 (1968); *Chem. Abstr.*, **69**, 18736d (1968).
61. L. L. Miller and A. K. Hoffmann, *J. Am. Chem. Soc.*, **89**, 593 (1967).
62. F. M. Beringer and R. A. Nathan, *J. Org. Chem.*, **34**, 685 (1969).
63. M. Schmeiser, K. Dakmen, and P. Sartori, *Chem. Ber.*, **103**, 307 (1970).
64. F. M. Beringer and R. A. Nathan, *J. Org. Chem.*, **35**, 2095 (1970).
65. Z. Jezic, U. S. Patent 3,622,586 (1971); *Chem. Abstr.*, **76**, 59460 (1972).
66. A. M. Nesmeyanov, *Bull. Acad. Sci. USSR, Cl. Sci. Chim.*, 239 (1945); *Chem. Abstr.*, **40**, 2122 (1946).
67. R. Kh. Freidlia, E. M. Brainina, and A. N. Nesmeyanov, *Bull. Acad. Sci. USSR Cl. Sci. Chim.*, 647 (1945); *Chem. Abstr.*, **40**, 4686 (1946).
68. F. M. Beringer and S. A. Galton, *J. Org. Chem.*, **30**, 1930 (1965).
69. A. M. Neseyanov, T. P. Tolstaya, N. F. Soklova, V. N. Varfolomessva, and A. V. Petrakov, *Dokl. Akad. Nauk SSSR*, **198**, 115 (1971); *Chem Abstr.*, **75**, 48576t (1971).
70. A. M. Nesmeyanov, T. P. Tolstaya, and A. V. Petrakov, *Dokl. Akad. Nauk SSSR*, **197**, 1337 (1971); *Chem. Abstr.*, **75**, 1403P2c (1971).

71. S. Kalhina and O. Nieland, *Z. Org. Khim.*, 7, 1606 (1971); *Chem. Abstr.*, **75**, 140102h (1971).
72. O. Neilands and G. Vanags, *Dokl. Akad. Nauk SSSR.*, **159**, 373 (1964); *Chem. Abstr.*, **62**, 6510C (1965).
73. Z. Jezic, 161st ACS National Meeting, Division of Medicinal Chemistry, Los Angeles, March 29-31, 1971.
74. Y. Yamada, K. Kashima, and M. Okawara, unpublished results.
75. Y. Yamada and M. Okawara, *Makromol. Chem.*, **152**, 153, 163 (1972).
76. J. M. Briody, *J. Chem. Soc., B,* 93 (1968).
77. D. L. LeCount and J. A. W. Reid, *J. Chem. Soc., C,* 1298 (1967).
78. W. V. Medlin *J. Am. Chem. Soc.*, **57**, 1026 (1935).
79. T. L. Khotsyanova, *Kristallografiya,* **2**, 51 (1957); *Chem. Abstr.*, **52**, 1704b (1958).
80. L. Irving, *J. Chem. Soc.*, **4498** (1962).
81. M. Arshad, M. A. A. Beg, and Y. Z. Abbasi, *Pak. J. Sci. Ind. Res.*, **12**, 12 (1969); *Chem. Abstr.*, **72**, 72804Z (1970).
82. F. M. Beringer and I. Lillien, *J. Am. Chem. Soc.*, **82**, 5153 (1960).
83. O. A. Ptitsyna, Y. V. Leuashova, M. E. Gurskii, and O. A. Reutov, *Izv. Akad. Nauk SSSR, Ser, Khim.*, **1970**, 197.
84. F. M. Beringer and S. A. Galton, *J. Org. Chem.*, **31**, 1648 (1966).
85. V. S. Petrosyan and O. A. Reutov, *Dokl. Akad. Nauk SSSR,* **175**, 613 (1967); *Chem. Abstr.*, **68**, 7940u (1968).
86. A. N. Nesmeyanov, N. M. Sergeev. Yu. A. Ustynyuk, T. P. Tolstaya, and I. N. Lisichkina, *Izv. Akad. Nauk SSSR, Ser. Khim.*, 154 (1970); *Chem. Abstr.*, **72**, 127069w (1970).
87. G. A. Olah, P. W. Westerman, E. G. Melby, and Y. K. Mo, *J. Am. Chem. Soc.*, **96**, 3565 (1974).
88. F. M. Beringer and I. Lillien, *J. Am. Chem. Soc.*, **82**, 5141 (1960).
89. O. A. Ptitsyna, G. G. Lyatiev, V. N. Krishchenko, and O. A. Reutov, *Izv. Akad. Nauk SSSR, Ser. Khim.*, 995 (1967); *Chem. Abstr.*, **67**, 111869t (1967).
90. G. Wittig and M. Rieber, *Ann. Chem.*, **562**, 187 (1949).
91. F. M. Beringer and P. S. Forgione, *Tetrahedron,* **19**, 739 (1963); F. M. Beringer and P. S. Forgione, *J. Org. Chem.*, **28**, 714 (1963); F. M. Beringer, P. S. Forgione and M. D. Yudis, *Tetrahedron,* **8**, 49 (1960); F. M. Beringer, S. A. Galton, and S. J. Huang, *J. Am. Chem. Soc.*, **84**, 2819 (1962).
92. K. G. Hampton, T. M. Harris, and C. R. Hauser, *Org. Synth.*, **51**, 128 (1971); K. G. Hampton, T. M. Harris, and C. R. Hauser, *J. Org. Chem.*, **29**, 3511 (1964).
93. F. M. Beringer, W. J. Daniel, S. A. Galton, and G. Rubin, *J. Org. Chem.*, **31**, 4315 (1966).
94. F. M. Beringer and L. L. Chang, *J. Org. Chem.*, **37**, 1516 (1972).
95. A. F. Lewit, N. N. Kalibabchuck, and I. P. Gragerov, *Dokl. Akad. Nauk SSSR,* **199**, 1325 (1971); *Chem. Abstr.*, **76**, 3131y (1972).
96. F. M. Beringer and S. A. Galton, *J. Org. Chem.*, **30**, 1930 (1965).
97. O. A. Ptitsyna, G. G. Lyatiev, and O. A. Reutov, *Izv. Akad. Nauk SSSR, Ser. Khim.*, 2125 (1968); *Chem. Abstr.*, **70**, 28504m (1969).
98. O. A. Ptitsyna, O. A. Reutov, and G. G. Lyatiev, *Zh. Org. Khim.*, **4**, 401 (1968); *Chem. Abstr.*, **68**, 104319f (1968).
99. O. A. Ptitsyna, O. A. Reutov, and G. G. Lyatiev, *Zh. Org. Khim.*, **5**, 708 (1969); *Chem. Abstr.*, **71**, 21766k (1969).
100. O. A. Reutov, O. A. Ptitsyna, and G. G. Lyatiev, *Izv. Akad. Nauk SSSR, Ser. Khim.*, **1967**, 631; *Chem. Abstr.*, **68**, 12234w (1968); O. A. Ptitsyna, G. G. Lyatiev, and O. A. Reutov, *Dokl. Akad. Nauk SSSR,* **182**, 119 (1968); *Chem. Abstr.*, **70**, 2900n (1969).

101. K. A. Bilevich, N. N. Bubnov, N. T. Ioffe, M. I. Kalinkin, O. Yu. Okhlobystin, P. V. Petrovskii, *Izv. Akad. Nauk SSSR, Ser. Khim.*, 1814 (1967); *Chem. Abstr.*, **76**, 3147h (1972).
102. O. A. Ptitsyna, G. G. Lyatiev, and O. A. Reutov, *Zh. Org. Khim.*, **5**, 974 (1969); *Chem. Abstr.*, **71**, 70211j (1969).
103. Parke, Davis & Co., Neth. Appl. B507,783 (1964); *Chem. Abstr.*, **64**, 19501d (1966).
104. F. M. Beringer, A. Brierley, M. Drexler, E. M. Gindler, and C. C. Lumpkin, *J. Am. Chem. Soc.*, **75**, 2708 (1953).
105. G. G. Lyatiev, O. A. Ptitsyna, and O. A. Reutov, *Zh. Org. Khim.*, **5**, 411 (1969); *Chem. Abstr.*, **71**, 2781v (1969).
106. O. A. Ptitsyna, G. G. Lyatiev, and O. A. Reutov, *Dokl. Akad. Nauk SSSR*, **157**, 364 (1964); *Chem. Abstr.*, **61**, 4417e (1964).
107. O. A. Ptitsyna, G. G. Lyatiev, and O. A. Reutov, *Zh. Org. Khim.*, **5**, 416 (1969); *Chem. Abstr.*, **71**, 2775w (1969).
108. A. Padwa, D. Eastman, and L. Hamilton, *J. Org. Chem.*, **33**, 1317 (1968).
109. O. A. Ptitsyna, M. E. Pudeeva, and O. A. Reutov, *Dokl. Akad. Nauk SSSR*, **165**, 838 (1965); *Chem. Abstr.*, **64**, 5129a (1966).
110. O. A. Ptitsyna, M. E. Pudeeva, and O. A. Reutov, *Dokl. Akad. Nauk SSSR*, **165**, 582 (1965); *Chem. Abstr.*, **64**, 19660h (1966).
111. O. A. Ptitsyna, M. E. Pudeeva, and O. A. Reutov, *Dokl. Akad. Nauk SSSR*, **168**, 595 (1966); *Chem. Abstr.*, **65**, 8712b (1966).
112. A. G. Varvoglis, *Tetrahedron Lett.*, 31 (1972).
113. A. N. Nesmeyanov, L. G. Makarova, and V. N. Vinogradova, *Izv. Akad. Nauk SSSR, Ser. Khim.*, 1966 (1969); *Chem. Abstr.*, **72**, 21758e (1970).
114. O. A. Ptitsyna, S. I. Orlov, and O. A. Reutov, *Vestn. Mosk. Univ., Ser. II, Khim.*, **21**, 105 (1966); *Chem. Abstr.*, **65**, 13755 (1966); O. A. Ptitsyna, S. I. Orlov, and O. A. Reutov, *Izv. Akad. Nauk SSSR, Ser. Khim.*, 1947 (1966); *Chem. Abstr.*, **66**, 75433q (1967).
115. T. V. Levashova, O. A. Ptitsyna, and O. A. Reutov, *Izv. Akad. Nauk SSSR, Ser. Khim.*, 1200 (1970); *Chem. Abstr.*, **73**, 65729w (1970).
116. O. A. Ptitsyna, T. V. Levashova, and O. A. Reutov, *Izv. Akad. Nauk SSSR, Ser. Khim.*, 165 (1968); *Chem. Abstr.*, **69**, 87140s (1968).
117. O. A. Ptitsyna, S. I. Orlov, M. M. Il'iva, and O. A. Reutov, *Dokl. Akad. Nauk SSSR*, **177**, 862 (1967); *Chem. Abstr.*, **68**, 69102f (1968).
118. J. A. Azoo, F. G. Coll, and J. Grimshaw, *J. Chem. Soc., C*, 2521 (1969).
119. A. N. Nesmeyanov and L. G. Makarova, *Izv. Akad. Nauk SSSR, Ser. Khim.*, 1966 (1969); *Chem. Abstr.*, **72**, 21758e (1970).
120. N. A. Nesmeyanov, S. T. Zhuzhlokova, and O. A. Reutov, *Izv. Akad. Nauk SSSR, Ser. Khim.*, 194 (1965); *Chem. Abstr.*, **62**, 11848c (1965).
121. P. B. Block, Jr. and D. H. Coy, *J. Chem. Soc., Perkin I*, 633 (1972).
122. S. B. Hamilton, Jr., and H. S. Blanchard, *J. Org. Chem.*, **35**, 3348 (1970).
123. E. J. Grubbs, R. J. Milligan, and M. H. Goodrow, *J. Org. Chem.*, **36**, 1780 (1971).
124. O. A. Ptitsyna, G. G. Lyatisev, and O. A. Reutov, *Dokl. Akad. Nauk SSSR*, **181**, 895 (1968); *Chem. Abstr.*, **69**, 96095h (1968).
125. F. M. Beringer and R. A. Falk, *J. Chem. Soc.*, 442 (1964).
126. F. M. Beringer, A. Brierley, M. Drexler, E. M. Grindler, and C. C. Lumpkin, *J. Am. Chem. Soc.*, **75**, 2708 (1953); F. M. Beringer, E. J. Geering, I. Kuntz, and M. Mausner, *J. Phys. Chem.*, **60**, 141 (1956).
127. F. M. Beringer, E. M. Gindler, M. Rapoport, and R. J. Taylor, *J. Am. Chem. Soc.*, **81**, 351 (1959); M. C. Caserio, D. L. Glusker, and J. D. Roberts, *ibid.*, **81**, 336 (1959).
128. F. M. Beringer and P. Bodlaender, *J. Org. Chem.*, **34**, 1981 (1969).
129. J. M. Davidson and G. Dyer, *J. Chem. Soc., A*, 1616 (1968).

130. D. J. LeCount and J. A. W. Reid, *J. Chem. Soc., C,* 1297 (1967).
131. Y. Yamada and M. Okawara, *Bull. Chem. Soc. Jap.*, **45**, 1860 (1972).
132. Y. Yamada and M. Okowara, *Bull. Chem. Soc. Jap.*, **45**, 2515 (1972).
133. A. N. Nesmeyanov, L. G. Makarova, and T. P. Tolstoya, *Tetrahedron,* **1**, 145 (1957).
134. W. H. Kirkle and G. F. Koser, *J. Am. Chem. Soc.*, **90**, 3598 (1968).
135. For a review see D. F. Banks, *Chem. Rev.*, **66**, 259 (1966).
136. L. Gershenfled and C. Kruse, *Am. J. Pharm.*, **121**, 343 (1969).
137. G. N. Pershin and N. S. Bodanova, *Farmakol. Toksikol.*, **25**, 209 (1962). *Chem. Abstr.*, 5715605C (1962).
138. G. N. Pershin, N. S. Bogadanova, K. I. Znaeva, and M. Ya. Kraft, *Farmakol. Toksifol.*, **24**, 609 (1961); *Chem. Abstr.*, **58**, 12885f (1963).
139. J. H. Holzaepfel, R. E. Greenbee, R. E. Wyand, and W. C. Ellis, Jr., *Fertility Sterility,* **10**, 272 (1959); *Chem. Abstr.*, **54**, 25577i (1960).
140. O. Neilands and B. Karele, *Zh. Org. Khim.*, **1**, 1854 (1965); *Chem. Abstr.*, **64**, 3396b (1966).
141. O. Neilands, M. Sile, and B. Karele, *Latv. PSR Zinat. Akad. Vestis, Khim. Ser.*, **1965**, 217. *Chem. Abstr.*, **63**, 13166f (1965).
142. O. Neilands and B. Karele, *Zh. Org. Khim.*, **2**, 488 (1966); *Chem. Abstr.*, **65**, 8869e (1966).
143. O. Neilands, *Zh. Org. Khim.*, **1**, 1858 (1965); *Chem. Abstr.*, **64**, 3396d (1966).
144. E. Gudriniece, O. Neilands, and G. Vanags, *Zh. Obsch. Khim.*, **27**, 2737 (1957); *Chem. Abstr.*, **52**, 7177 (1958).
145. O. Neilands and D. Neimanis, *Zh. Org. Khim.*, **6**, 2509 (1970); *Chem. Abstr.*, **74**, 64235r (1971).
146. O. Neilands and B. Karele, *Zh. Org. Khim.*, 7, 1611 (1971); *Chem. Abstr.*, **75**, 40768e (1971).
147. B. Karele and O. Neilands, *Zh. Org. Khim.*, **2**, 1680 (1966); *Chem. Abstr.*, **66**, 64882j (1967).
148. D. Neimanis and O. Neilands, *Zh. Org. Khim.*, **6**, 633 (1970); *Chem. Abstr.*, **72**, 64882j (1967).
149.a. D. Neimanis and O. Neilands, *Zh. Org. Khim.*, **6**, 1011 (1970); *Chem. Abstr.*, **73**, 34979y (1970
149.b. D. Prikule and O. Neilands, *Zh. Org. Khim.*, 7, 2441 (1971); *Chem. Abstr.*, **76**, 58766q (1971).
150. Y. Hayashi, T. Okada, and M. Kawanisi, *Bull. Chem. Soc. Jap.*, **43**, 2506 (1970).
151. W. A. Sheppard and O. W. Webster, *J. Amer. Chem. Soc.*, **95**, 2695 (1973).
152. Y. Yamada and M. Okawara, unpublished results.
153.a. E. D. Hughes, C. K. Ingold et al., *J. Chem. Soc.*, 1196 (1937).
153.b. S. Winstein and R. Boschan, *J. Am. Chem. Soc.*, **72**, 4669 (1950), and previous papers
154. G. A. Olah, D. A. Beal, and P. W. Westerman, *J. Am. Chem. Soc.*, **95**, 3387 (1973).
155. G. A. Olah, Y. K. Mo, and Y. Halpern, *J. Org. Chem.*, **37**, 1169 (1972).
156. D. T. Clark, in *Special Lectures of XXIII International Congress of Pure and Applied Chemistry, Boston, 1971,* Vol. I, Butterworth, London, pp. 31.
157. J. M. Bollinger, J. M. Brinich, and G. A. Olah, *J. Am. Chem. Soc.*, **92**, 4025 (1970).
158. G. A. Olah and A. M. White, *J. Am. Chem. Soc.*, **91**, 5801 (1969).
159. T. F. Page, T. Alger, and D. M. Grant, *J. Am. Chem. Soc.*, 87, 5333 (1965).
160. J. B. Stothers, *Carbon-13 Nmr Spectroscopy,* Academic Press, New York, 1972, p. 55.
161. P. E. Peterson and B. R. Bonazza, *J. Org. Chem.*, **38**, 1010 (1973).
162. G. A. Olah, C. L. Jeuell, and A. M. White, *J. Am. Chem. Soc.*, **91**, 3961 (1969).
163. G. A. Olah and J. M. Bollinger, *J. Am. Chem. Soc.*, **90**, 6082 (1968).

164.a. D. V. Banthorpe, *Chem. Rev.*, **70**, 295 (1970).

164.b. P. B. D. de la Mare and R. Bolton, *Electrophilic Additions to Unsaturated Systems*, Elsevier, Amsterdam, 1966.

164.c. I. V. Bodrikov and Z. S. Smolyan, *Russ. Chem. Rev.*, **35**, 374 (1966).

165. G. A. Olah, P. Schilling, P. W. westerman, and H. C. Lin, *J. Am. Chem. Soc.*, **96**, 3581 (1974).

166. J. Strating, J. H. Wieringa, and H. Wynberg, *Chem. Commun.*, 907 (1969).

167. J. H. Wieringa, J. Strating, and H. Wynberg, *Tetrahedron Lett.*, 4579 (1970).

168. R. E. Glick, Ph.D. Thesis, University of California at Los Angeles, Los Angeles, Calif., 1954.

169. P. E. Peterson and G. Allen, *J. Am. Chem. Soc.*, **85**, 3608 (1963).

170. P. E. Peterson and E. V. P. Tao, *J. Am. Chem. Soc.*, **86**, 4503 (1964).

171. (a) P. E. Peterson and J. E. Duddey, *ibid.*, **88**, 4990 (1966) and (b) P. E. Peterson and R. J. Bopp, Abstracts, 152nd National Meeting of the American Chemical Society, New York, N.Y., Sept. 12-16, 1966, Paper S3.

172. P. E. Peterson, H. J. Bopp, D. M. Chevli, E. Curran, D. Dillard, and R. J. Kamat, *ibid.*, **89**, 5902 (1967).

173. G. A. Olah, J. M. Bollinger, and J. M. Brinich, *J. Am. Chem. Soc.*, **90**, 6988 (1968).

174. G. A. Olah, J. M. Bollinger, Y. K. Mo, and J. M. Brinich, *J. Am. Chem. Soc.*, **94**, 1164 (1972).

175. G. A. Olah, Y. K. Mo, E. G. Melby, and H. Lin, *J. Org. Chem.*, **38**, 367 (1973).

176. P. E. Peterson, P. R. Clifford and F. J. Slama, *J. Am. Chem. Soc.*, **92**, 2840 (1970).

177. P. E. Peterson and B. R. Bonazza, *J. Am. Chem. Soc.*, **94**, 5017 (1972).

178. P. M. Henrichs, and P. E. Peterson, *J. Am. Chem. Soc.*, **95**, 7449 (1973).

179. G. A. Olah, J. R. DeMember, and R. H. Schlosberg, *J. Am. Chem. Soc.*, **91**, 2112 (1969).

180. B. R. Bonazza and P. E. Peterson, *J. Org. Chem.*, **38**, 1010 (1973).

181. J. W. Larsen, and A. V. Metzner, *J. Am. Chem. Soc.*, **94**, 1614 (1972).

182. S. W. Benson, F. R. Cruickshank, D. M. Golden, G. R. Haugen, H. E. O'Neal, A. S. Rodgers, R. Shaw, and R. Walsh, *Chem. Rev.*, **69**, 279 (1969).

183. D. V. Banthorpe, *Chem. Rev.*, **70**, 295 (1970).

184. See G. A. Olah, *J. Am. Chem. Soc.*, **94**, 808 (1972) for a discussion of π- and σ-type intermediates.

185. P. R. Jones, *J. Org. Chem.*, 37, 1886 (1972).

186. For a comprehensive review, see G. A. Olah, *Acc. Chem. Res.*, **4**, 240 (1971) and references therein.

187. G. A. Olah, *Angew. Chem. Int. Ed.*, **12**, 173 (1973); G. A. Olah *Carbocations and Electrophilic Reactions*, Verlag-Chemie-Wiley, 1973.

188. F. R. Mayo and J. J. Katz, *J. Am. Chem. Soc.*, **69**, 1339 (1947).

189. C. G. Gebelein and G. D. Frederick, *J. Org. Chem.*, **37**, 2211 (1972).

190. R. E. Buckles and J. P. Yuk, *J. Am. Chem. Soc.*, **75**, 5048 (1953).

191. J. E. Dubois and F. Garnier, *Spectrochim. Acta, Part A*, **23**, 2279 (1967).

192. V. A. Lishnevskii and G. B. Sergeev, *Kinet. Katal.*, **5**, 407 (1964).

193. J. E. Dubois and G. Nouvier, *Tetrahedron Lett.*, 1325 (1963).

194. G. A. Olah and T. R. Hockswender, Jr., *J. Am. Chem. Soc.*, **96**, 3574 (1974).

195. M. A. Muhs and F. T. Weiss, *J. Am. Chem. Soc.*, **84**, 4697 (1962).

196. L. J. Andrews and R. M. Keefer, *Molecular Complexes in Organic Chemistry*, Holden-Day, San Francisco, 1964, p. 88.

197. R. J. Cvetanovic, F. J. Duncan, W. E. Falconer, and W. A. Sunder, *J. Am. Chem. Soc.*, 88, 1602 (1966).

198. R. D. Bach and H. F. Henneike, *J. Am. Chem. Soc.*, **92**, 5589 (1970).
199. H. C. Brown and P. J. Geoghegan, Jr., *J. Org. Chem.*, **37**, 1937 (1972).
200. R. E. Erickson and R. L. Clark, *Tetrahedron Lett.*, 5997 (1969).
201. J. A. Pincock and K. Yates, *Can. J. Chem.*, **48**, 3332 (1970).
202. R. W. Alder, R. Baker, and J. M. Brown, *Mechanism in Organic Chemistry*, Wiley-Interscience, New York, 1971, p. 307.
203. R. T. Morrison and R. N. Boyd, *Organic Chemistry*, Allyn and Bacon, Boston, 1966, p. 235ff.
204. G. A. Olah, unpublished results.
205. G. A. Olah and S. J. Kuhn, *J. Org. Chem.*, **29**, 2317 (1964).
206. I. L. Reich and H. J. Reich, *J. Am. Chem. Soc.*, **96**, 2654 (1974).
207. F. M. Beringer, P. Ganis, G. Avitabile, and H. Jaffe, *J. Org. Chem.*, **37**, 879 (1972).
208. F. M. Beringer and R. A. Nathan, *J. Org. Chem.*, **34**, 685 (1969).
209. W. N. Cannon, U. S. Patent 3,264,355 (1964); *Chem. Abstr.*, **65**, 13664a (1966).
210. L. Mascarelli, *Gazz. Chim. Ital.*, **43**, 26 (1913).
211. R. B. Sandin, *J. Org. Chem.*, **34**, 456 (1969).
212. W. Baker, J. W. Barton, and J. F. W. McOmine, *J. Chem. Soc.*, 2658 (1958).
213. H. Heaney and P. Lees, *Tetrahedron*, **24**, 3717 (1968).
214. R. B. Sandin and A. S. Hay, *J. Am. Chem. Soc.*, **74**, 274 (1952).
215. K. Clup, *Chem. Ber.*, 88, 268 (1955).
216. T. Sato, S. Shimada, K. Shimizu, and K. Hata, *Bull. Chem. Soc. Jap.*, **43**, 1918 (1970).
217. F. M. Beringer and L. L. Chang, *J. Org. Chem.*, **36**, 4055 (1971).
218. R. C. Fuson and R. L. Albright, *J. Am. Chem. Soc.*, **81**, 487 (1959).

Author Index

Abbasi, Y., 2, 172
Albright, R. L., 176
Alder, R. W., 176
Alekseeva, L. A., 171
Alger, T., 174
Allen, G., 175
Allinger, N. L., 169
Allkin, D. R., 170
Andrews, L. J., 175
Arshad, M., 64, 172
Avitabile, G., 176
Azoo, J. A., 173

Bach, R. D., 176
Bachofner, H. F., 171
Baker, R., 176
Baker, W., 176
Banks, D. F., 170, 174
Banthrope, D. V., 175
Barton, J. W., 176
Baumgarten, E., 32, 171
Beal, D. A., 174
Beauchamp, J. L., 30, 171
Beg, A. A., 172
Benson, S. W., 148, 175
Beringer, F. M., 1, 57, 58, 67, 74, 83, 158, 159, 171, 172, 173, 176
Bilevich, K. A., 173
Blanchard, H. S., 173
Block, P. B., Jr., 173
Bodanova, N. S., 174
Bodlaender, P., 173
Bodrikov, I. V., 175
Bollinger, J. M., 2, 98, 108, 110, 170, 174, 175
Bolton, R., 175
Bonazza, B. R., 129, 146, 157, 170, 174, 175
Bond, K. C., 32, 171
Bopp, R. J., 175
Boron, B. F., 170
Boschan, R., 174
Boyd, R. N., 176
Boyton, H. G., 171
Brainiana, E. M., 171
Brierley, A., 173
Brinich, J. M., 170, 174, 175
Briody, J. M., 172
Brown, H. C., 32, 171, 175, 176
Brown, J. M., 176
Bubnov, N. N., 173
Buckles, R. E., 175

Cannon, W. N., 176
Capon, B., 169
Caserio, M. C., 173
Chang, L. L., 172, 176
Chevli, D. M., 175
Clark, D. T., 109, 174
Clark, R. L., 176
Clifford, P. R., 175
Clup, K., 176
Coll, F. G., 173
Comisarow, M. B., 171
Commeyras, A., 171
Conway, D. C., 171
Coward, J., 32, 171
Coy, D. H., 173
Cruickshank, F. R., 175
Curran, E., 175
Cvetanovic, R. J., 175

Dalling, D. K., 170
Daniel, W. J., 172
Davidson, J. M., 173
DeHaan, H., 32, 171
Dehn, J. W., Jr., 171
de la Mare, P. B. D., 175
DeMember, J. R., 2, 8, 15, 169, 171, 175
Dence, T. B., 171
Dillard, D., 175
Doering, W. V. E., 171
Drexler, M., 171, 173
Dubois, J. E., 149, 153, 175
Duddey, J. E., 175
Duncan, F. J., 175
Dyer, G., 173

Earhart, H. W., 171
Eastman, D., 173
Ebel, E. L., 169
Edwards, W. E., 171

Ellis, W. C., Jr., 174
Erickson, R. E., 176
Exner, J. H., 141, 170

Falconer, W. E., 175
Falk, R. A., 171, 173
Fay, R. C., 169
Fort, R. C., Jr., 171
Frederick, G. D., 175
Freidlia, R. Kh., 171
Fuson, R. C., 176

Galton, S. A., 171, 172
Ganis, P., 176
Gebelein, C. G., 175
Geering, E. J., 173
Geoghegan, P. J., 176
Gershenfeld, L., 174
Gibley, M. G., 171
Gindler, E. M., 171, 173
Glick, R. E., 175
Golden, D. M., 175
Goodrow, M. H., 173
Gould, E., 2, 170
Gragerov, I. P., 172
Grant, D. M., 16, 170, 174
Grebbs, E. J., 173
Greenbee, R. E., 174
Grimshaw, J., 173
Gudriniece, E., 174
Guerskii, M. E., 171
Gueskev, D. L., 173

Halpern, Y., 171, 174
Hamilton, L., 173
Hamilton, S. B., Jr., 173
Hampton, K. G., 172
Hantsch, A., 4
Harris, J. M., 172
Hartman, C., 1, 54, 169
Hata, K., 176
Haugen, G. R., 175
Hauser, C. R., 69, 172
Hay, A. S., 168, 176
Hayashi, Y., 92, 174
Heaney, H., 165, 168, 176
Hendra, P. J., 170
Hendrics, P. M., 170
Henneike, H. F., 176
Henrics, P. M., 133, 175
Herzberg, G., 170
Hockswender, T. R., Jr., 149, 175
Hoffmann, A. K., 57, 171
Holzaepfel, J. H., 174
Huang, S. J., 172
Hughes, E. D., 98, 174
Hwang, W. K., 171

Il'iva, M. M., 173
Ingold, C. K., 98, 174
Ioffe, N. T., 173
Irving, L., 172

Jaffe, H., 176
Jeuell, C. L., 174
Jezic, Z., 62, 171, 172
Jones, P. R., 175

Kalhina, S., 172
Kalibabchuck, N. N., 172
Kalinkin, M. I., 173
Kamat, R. J., 175
Karele, B., 174
Karniol, M., 171
Kartz, J. J., 175
Kashima, K., 172
Kawanisi, M., 174
Keefer, R. M., 175
Kennedy, J. P., 31, 171
Kershner, L. D., 170
Khotsyanova, T. L., 64, 172
Kimball, G. E., 1, 98, 169
Kinsella, E., 32, 171
Kirkle, W. H., 174
Koser, G. F., 174
Kovsky, T. E., 170
Kraft, M. Ya., 174
Krishchenko, V. N., 172
Kruse, C., 174
Kuhn, S. J., 176
Kuntz, I., 173

Laber, G., 171
Larsen, J. W., 147, 148, 152, 175
LeCount, D. L., 63, 172, 174
Lees, P., 176
Leff, M., 171
Levashova, T. V., 172, 173
Lewitt, A. F., 172
Liang, G., 140, 170, 171

Lillien, I., 171, 172
Lin, H. C., 170, 175
Lisichkina, I. N., 171, 172
Lucas, H. J., 1, 98, 170
Lumpkin, C. C., 171, 173
Lyalin, V. V., 171
Lyatiev, G. G., 172, 173

Makarova, L. G., 169, 173, 174
Mascarelli, L., 176
Masson, I., 171
Masulio, G., 171
Mausner, M., 171, 173
Mayo, F. R., 175
Medlin, W. V., 63, 172
Meerwein, H., 21, 170
Melby, E. G., 39, 170, 171, 172, 175
Meomine, J. F. W., 176
Metznev, A. V., 147, 175
Meyer, V., 1, 54, 169
Miller, L. L., 57, 171
Mo, Y. K., 170, 172, 174, 175
Morrison, R. T., 176
Muhs, M. A., 175
Muller, D., 170
Mulligan, R. J., 173

Nathan, R. A., 159, 171, 176
Neilands, O., 90, 172, 174
Neimanis, D., 174
Nesmeyanov, A. N., 1, 56, 60, 67, 87, 169, 171, 172, 173, 174
Nouvier, G., 175

Okada, T., 174
Okawara, M., 63, 85, 86, 94, 172, 174
Okhlobystin, O. Yu., 173
Olah, G. A., 2, 4, 8, 15, 39, 67, 98, 108, 110, 111, 113, 120, 125, 134, 135, 144, 149, 161, 169, 170, 171, 172, 174, 175, 176
Olah, J. A., 170
O'Neal, H. E., 175
Orda, V. V., 171
Orlov, S. I., 173

Padwa, A., 173
Page, T. F., 174
Paul, E. G., 170
Perkampus, H. H., 32, 171
Pershin, G. N., 174
Peterson, P. E., 2, 4, 111, 125, 129, 133, 134, 135, 142, 146, 156, 157, 170, 174, 175
Petrakov, A. V., 171
Petrosyan, V. S., 66, 172
Petrovskii, P. V., 173
Pincock, J. A., 154, 176
Plesch, P. H., 31, 171
Prikule, D., 174
Ptitsyana, O. A., 172, 173
Pudeeva, M. E., 173

Race, E., 171
Rapport, M., 173
Reich, H. J., 176
Reich, I. L., 176
Reid, J. A. W., 174
Renner, R., 170
Reutov, O. A., 172, 173
Rice, B., 32, 171
Rieber, M., 172
Roberts, I., 1, 98, 169
Roberts, J. D., 171, 173
Rodgers, A. S., 175
Rubin, G., 172

Sandin, R. B., 165, 168, 170, 176
Sato, T., 176
Saunders, M., 171
Schilling, P., 170, 175
Schleyer, P. v. R., 171
Schlosberg, R. H., 171, 175
Sergeev, N. M., 172
Shaw, R., 175
Sheppard, W. A., 174
Shimada, S., 176
Shimiuz, K., 176
Siebert, H., 170
Sile, M., 174
Slama, F. J., 175
Smolyan, Z. S., 175
Soklova, N. F., 171
Spear, R. J., 170
Staley, R. H., 171
Staral, J., 144, 170
Stothers, J. B., 170, 174
Strating, J., 175
Summer, E., 171
Sunder, W. A., 175

Svoboda, J. J., 170

Tao, E. V. P., 175
Taylor, R. J., 173
Tolstaya, T. P., 169, 171, 172, 174
Traynham, J. C., 170

Ustynyuk, Yu. A., 172

Vanags, G., 174
Varfolomeeva, V. N., 171
Varvoglis, A. G., 173
Vinogradova, V. N., 173
Vogel-Hogler, R., 170

Wadley, E. F., 171
Walker, D. G., 32, 171
Wallace, W. J., 32, 171
Walsh, R., 175
Watts, W. E., 171
Webster, O. W., 174
Weiss, F. T., 175
Westerman, P. W., 170, 172, 174, 175
White, A. M., 113, 171, 174
Wieringa, J. H., 120, 175
Wieting, R. D., 171
Winicov, M., 171
Winstein, S., 1, 98, 170, 174
Wittig, G., 172
Wyand, R. E., 174
Wynberg, H., 120, 175

Yagupoliskii, L. M., 171
Yamada, Y., 63, 85, 86, 94, 161, 170
Yates, K., 154, 176
Yudis, M. D., 172
Yuk, J. P., 175

Zhuzhlokova, S. T., 173
Znaeva, K. I., 174

Subject Index

Acetic acid, alkylation of, 156
Acetolysis, of 4-bromo-1-butyl tosylate, 125
  of 4-iodo-1-butyl tosylate, 125
Acetone, alkylation of, 33
Acetonitrile, alkylation of, 156
Acetoxonium ions, 110
N-Acetyl-3,4-diiodo-L-phenylalanine ethyl ester, 81
Acidic halonium ions, 2, 4
Acyclic halonium ions, 2, 5
Adamantyl bromide, halide transfer of, 28
1-Adamantyl cation, 27, 30
Adamantyl chloride, halide transfer of, 28
Adamantyl halides, 27
Adamantylideneadamantane, 120, 150
  bromine complex, 122
Aldehydes, alkylation of, 21
Aliphatic amines, reaction with diphenyliodonium ions, 74
Alkali iodates, 44
Alkenehalonium ion, 148
Alkenes, 149
  alkylation of, 31
  bromine-π complex, 149
    role of σ and π complexes in the addition of halogens, 148
  differentiation of π and σ bonded complexes of, 120
  halogen complexes of, 120
  polymerization of, 31, 131
Alkylarylhalonium ions, 1, 2, 39
Alkylating ability, 45
  preparation and NMR study, 39
  relative stability, 45
  their role in Fridel-Crafts reactions, 45
Alkylating agents, 19, 169
Alkylation, 8, 9
  of aldehydes, 21
  of alkenes, 31
  of amino compounds, 21
  of aromatics, 19
  of carboxylic acids, 21
  of dihaloarenes, 45
  of ether, 21
  of hetero organic compounds, 21
  of ketones, 21
  of nitro compounds, 21
  of sulfur compounds, 21
  of water, 21
Alkyl carbenium ions, 9
  intermolecular exchange reactions of, 23
Alkylenehalonium ions, 2, 4
Alkyl fluorides, 33
Alkyl fluoroantimonates, 8
Alkyl halides, intermolecular exchange reactions of, 23
  self-condensation of, 6
Alkylhalonium ylides, 90
Alkylhydridohalonium ions, 6
Alkyl iodides, 19
Alkyl norbornyl halonium ions, 30
Alkynes, bromination of, 154
  role of σ and π complexes in the addition of halogens, 148
Alkyl phosphates, reaction with diaryliodonium ions, 78
Alkyl phosphines, reaction with diaryliodonium ions, 78
Aluminum bromide, complex formation with alkyl bromides, 31
Amines, reaction with diaryliodonium ions, 72
Amino compounds, alkylation of, 21
2-Amino-2′-halobiphenyls, 168
Ammonium ions, 169
Aniline, reaction with diphenyliodonium ion, 72
Anion exchange reactions, 39
Anionic reagents, reaction with diphenylhalonium ions, 68
Aromatics, alkylation of, 19
Arylated amines, 74
Aryl cations, 57
Aryl(*trans*-chlorovinyl)iodonium ion, 60
Aryldiazonium salts, 56
Arylhalonium ions, 4
Arylhalonium ylides, 90
Arylhydridohalonium ions, 6
Aryliodonio-5,5-dimethylcyclohexane-1,3-dione, 61
Aryliodoso acetate, 61, 62
Aryliodoso chloride, 56, 60

Aryl 2-thienyliodonium ion, 87
Aziridines, reaction with diaryliodonium ions, 75

Benzchlorophenium ions, 161
Benzene, alkylation of, 19
  diaryliodonium ion formation with, 62, 63, 73
Benzenium ions, 6
Benziodophenium ions, 158
Benziodophenium salts, 159
Benzoic acid, formation from diaryliodonium ions, 84
Benzophenone oxime, reaction with diaryliodonium ions, 82
α-Benzoyl acetate, 93
Benzoyl cations, 44
Bicyclic halonium ions, 4, 144

cis-1,2-Bis(chloromethyl)-cyclohexane, ionization of, 146
Bis(4-chlorophenyl)phenyliodine, 69
Bis(*trans*-2-chlorovinyl)iodonium chloride, 59
Bis(3,4-dichlorophenyl)iodonium chloride, 89
Bis(2,4-dichlorophenyl)iodonium sulfate, 89
4,4′-Bis(dimethylamino)diphenyl iodonium chloride, 70
Bis(N,N-dimethyl)phenylene diamine, reaction with diphenyliodonium ion, 73
3,3-Bis(halomethyl)trimethylenebromonium ions, 141
Bisiodonium ions, 163
Bridged bromonium ions, 1
Bridged fluoronium ions, 110
Bromination, of alkynes, 154
  of cyclohexene, 153
  of cyclopentene, 153
  of norbornene, 153
Bromobenzene, alkylation of, 42
  protonation of, 6
  ylide formation of, 93
4-bromo-1-butyl tosylate, acetolysis of, 125
1-Bromo-4-chloropentane, ionization of, 128
β-Bromocumyl ions, 116
4-Bromocyclopentene, 144
4-Bromo-2,6-dimethylphenol, 82
erythro-dl-2-Bromo-3-fluorobutane, 103
threo-dl-2-Bromo-3-fluorobutane, 103
Bromomethane, protonation of, 6
2-Bromonia[3,1,0] bicyclohexane, 4
2-Bromonia[3,1,0] cyclohexane, 144
Bromonium ions, 1
  bridged, 1
Bromonium ylides, 93
N-Bromosuccinimide, 121
3-Bromo-tetramethylenebromonium ion, 134
Bromoxylenes, formation from methyl, phenylbromonium ion, 45
1-Butene, alkylation of, 31
2-Butene, alkylation of, 31
Butyl, alkylchloronium ions, 135
tert-Butyl cation, 30, 148
1-Butylenebromonium ion, 112
1-Butylenechloronium ion, 112
1-Butyleneiodonium ion, 128
3-Butyl-2-phenylbenziodophenium chloride, 158
3-Butyl-2-phenbenziodophenium iodide, 159
2-Butyltetramethylene bromonium ion, 131

Carbenes, preparation by halonium ylides, 169
Carbenium ions, 1, 98
Carbocations, 149
Carbocation stabilities, determination of, 28
Carboxylic acids, alkylation of, 21
Carbon monoxide, reaction with diphenyliodonium ion, 84
Carbon-13 nmr studies, of ethylenehalonium ions, 113
  of tetramethylenehalonium ions, 135
Cationic polymerization, initiated by dialkylhalonium ions, 31
Charge transfer complexes, in halogen additions, 149
Chemical reactivity, 68
  of diiodonium ions, 68
  of halonium ions, 147
  of polyiodonium ions, 68
1,4-Chlorine participation, 125
ω-Chloroalkylcarbenium ion, 139
δ-Chloroalkyl tosylates, 125

Chlorobenzene, formation in diaryliodonium chloride decomposition, 69, 84, 85
protonation of, 6
ylideformation of, 93
Chlorocarbenium ion, 110
1-Chloro-1-(o-chlorophenyl)2-fluoropropene-1, 162
1-Chloro-1-cyclopentyl cation, 145
o-(β-Chloroethynyl)chlorobenzene, 161
1-Chloro-1-fluoro-2-(o-chlorophenyl)-pentene-1, 162
1-Chloro-1-fluoro-2-(o-chlorophenyl)-propene-1, 162
1-Chloro-2-fluoroethane, 98
4-Chloro-1-fluoro-4-methyl pentane, 130
trans-β-Chloroiodosochloroethylene, 59, 60
Chloromethane, protolytic behavior of, 6
3-Chloronia[4,3,0] bicyclononane, 4, 146
Chloronium ylides, 93
5-Chloro-1-pentene, 128
5-Chloro-1-pentyne, 135
5-Chloro-2-pentyl trifluoroacetate, 128
Chlorophenium ions, 158
3-Chlorotetramethylenechloronium ion, 134
Chlorotoluenes, 45
trans-Chlorovinyliodoso dichloride, 57, 158
trans-Chlorovinylmercuric chloride, 59, 60
Chromium V chloride, catalysis of decomposition of nalonium ions, 84
Competitive exchange, determination of relative carbocation stabilities through, 28
σ-Complexes, 149
π-Complexes, 149
Cyanogen bromide, 121, 120
Cyanogen iodide, 120
Cyclic halonium ions, 2, 97
Cyclic iodonium ions, 56
Cyclohexene, bromination of, 153
bromine transfer to, 122
Cyclopentene, bromination of, 153
Cyclopentenebromonium ion, 144
Cyclopentenechloronium ion, 145
Cyclopentenyl cation, 144
Cyclopentyl cation, 30
Cyclopropyl halides, protolytic behavior of, 118
Cyclopropyl lithium, 39
Cyclopropyl, phenyliodonium ion, 39
Deprotonation, 24
Diadmantylhalonium ions, 27
Dialkyl alkylenedibromonium ions, 38
Dialkyl alkylenedichloronium ions, 38
Dialkyl alkylenedihalonium ions, 9, 38
Dialkyl alkylenediiodonium ions, 38
Dialkylhalonium ions, 1, 2, 8, 54
alkylation and polymerization of alkenes by, 31
chemical reactivity, 19
determination of relative carbocation stabilities through competitive exchange of, 28
intermolecular exchange reactions of, 23
NMR spectroscopic studies, 15
pmr, 15
carbon-13 NMR, 15
preparation and isolation, 8
Raman and IR spectroscopic studies, 18
role of, in Friedel-Crafts reactions, 31
Dialkylphenylenedihalonium ions, 45
2,2′-Diaminobiphenyl, 164
Di-*tert*-amylhalonium ion, 27
Diarylamine, 74
Diarylbromonium ions, 87
Diaryl-*trans*-β-chloroethenyliodine, 58
Diarylchloronium ions, 87
Diarylhalonium ions, 2, 54
Diaryliodonium bisulfates, 55
Diaryliodonium chlorides, 55
Diaryliodonium compounds, 1
Diaryliodonium hydroxides, 80
Diaryliodonium ions, 2, 45, 65, 79, 89
biological activity of, 89
chemical reactivity of, 68
IR spectra, 64
kinetics of formation, 63
preparation of, 54
reactions with alkyl phosphates, 78
alkyl phosphites, 78
anionoic reagents, 69
aziridies, 75
cumene, 72
inorganic nucleophiles, 82
mercury, 80
O-donor substrates, 80
phosphines, 77
structural aspects of, 63
UV spectra, 64
X-ray studies, 63

Diaryliodonium nitrates, 55
Diazodicyanoimidazole, 93, 94
Dibenzhalophenium ions, 163
Dibenziodophenium ion, 70, 159, 163, 165
Dibenziodophenium sulfate, 167
Dibenzoylmethane, 92
p-Dibromobenzene, alkylation of, 45
dl-2,3-Dibromobutane, 103, 104, 115
meso-2,3-Dibromobutane, 103
trans-1,2-Dibromocyclopentane, 144
2,5-Dibromohexanes, 130
1,5-Dibromopentane, 128, 142
Dibromopropanes, 53
3,5-Di-*tert*-butylbenzene-1,4-diazooxide, 88
Di-*tert*-butylhalonium ions, 27
1,1-Dichloro-2-(2-biphenyl)propane-1, 162
Dichlorobutane, meso-2,3-, 106
  dl-, 106
2,2-Dichloro-3,3-dimethylbutane, 110
2,5-Dichloro-2,5-dimethylhexane, 133
4,4′-Dichlorodiphenyliodonium chloride, 69
1,2-Dichloroethane, 98
o-(β,β-Dichloroethenyl)phenyldiazonium fluorophosphates, 161
Dichlorohexanes, 130, 131
1,4-Dichloro-2-methylbutane, 129
2,5-Dichloro-2-methylhexane, 133
2,5-Dichloro-2-methylpentane, 130
1,5-Dichloropentane, 128
1,1-Dichloropropane, 102
1,2-Dichloropropane, 102
1,3-Dichloropropane, 102
1,1-Dichloro-2-(o-fluorophenyl)propene-1, 162
cis-4,5-Dichlorotetramethylene-chloronium ion, 134
Dicyanoimidazoleiodonium ylide, 94
Diethylamine, reaction with diaryliodonium ions, 75
N,N-Diethylaniline, reaction with diaryliodonium ions, 75
Diethylhalonium ions, 21
N,N-Diethyl-*m*-nitroaniline, 75
Diethyl phenylphosphonate, 78
1,4-Difluorobutane, 128
1,1-Difluoroethane, 101
2,2′-Difurfuryliodonium ion, 59
Dihaloarenes, alkylation of, 45
1,2-Dihalobutanes, 125
1,3-Dihalobutanes, 125
1,4-Dihalobutanes, 9, 125
1,3-Dihalo-2-methylpropanes, 140
1,5-Dihalopentanes, 142
1,3-Dihalopropanes, 101, 140
Diiodobenzene, dialkylation of, 45, 53
p-Diiodobenzene, 78, 79
2,2′-Diiodobiphenyl, 167
Diiodomethane, dialkylation of, 53
Diiodonium ions, 62
  chemical reactivity, 68
  degree of dissociation, 68
  kinetics of formation, 63
  IR spectra, 64
  NMR spectra, 66
  structural aspects, 63
  substituent effects in, 67
  UV spectra, 64
  X-ray studies, 63
2,2′-Diiodosodichlorobiphenyl, 163
2-Diiodo-2′-phenylbiphenyl, 166
Diisopropylbromonium ion, 24
Diisopropylchloronium ion, 9, 23, 24
Diisopropyliodonium ion, 24, 26
2,2′-Dilithiobiphenyl, reaction with diphenyliodonium ion, 70
1,4-Dilithio-1,2,3,4-tetraphenyl-1,3-butadiene, in preparation of tetraphenyliodophenium ion, 158
Dimedone, phenylation of, 69
Di(4-methoxy-3,4-dimethylphenyl)-iodonium iodide, 81
Dimethylamine, phenylation of, 74
4-Dimethylaminophenyl lithium, reaction with diphenyliodonium ion, 70
2,3-Dimethylaniline, phenylation of, 74
Dimethylbromonium ion, 6
Dimethylbromonium fluoroantimonate, 9, 31, 90
2,3-Dimethyl-2-butene, 120
2,3-Dimethyl-2-chlorobutane, 24
Dimethylchlorocarbenium ion, 9, 24
Dimethyl-β-chloroethylcarbenium ion, 129
Dimethylchloronium ion, 6, 9
5,5-Dimethylcyclohexane-1,3-dione, 61
2,3-Dimethyl-2,3-dibromobutene, 148
Dimethyl ether, alkylation of, 156
cis-1,2-Dimethylethylenebromonium ion, 145
2,2-Dimethylethylenebromonium ion, 104, 116, 118

2,3-Dimethylethylenebromonium ion, 115
Dimethylethylenehalonium ions, 106
trans-1,2-Dimethylethylenehalonium ion, 105
2,2-Dimethylethylenehalonium ions, 106
2,3-Dimethylethylenehalonium ions, 103
1,1-Dimethylethyleneiodonium ion, 107
1,2-Dimethylethyleneiodonium ion, 115
2,2-Dimethylethyleneiodonium ion, 106
2,3-Dimethylethyleneiodonium ion, 106, 115
Dimethyl-β-haloethylcarbenium ions, 141
Dimethylhalonium fluoroantimonates, 9
Dimethylhalonium ions, 8, 18, 19, 90
Dimethyliodonium fluoroantimonate, 9, 90
Dimethyliodonium ion, 6
Dimethylsulfide, model for dimethylhalonium ions, 19
Dimethylsulfoxide, reaction with halonium ylides, 95
1,1-Dimethyltetramethylenebromonium ion, 139
1,4-Dimethyltetramethylenebromonium ion, 138
2,2-Dimethyltetramethylenebromonium ion, 138
2,5-Dimethyltetramethylenebromonium ion, 130
1,1-Dimethyltetramethylenechloronium ion, 139
2,2-Dimethyltetramethylenechloronium ion, 130, 131, 138
2,2-Dimethyltetramethylenefluoronium ion, 132
2,5-Dimethyltetramethylenehalonium ion, 129
2,2-Dimethyltetramethyleneiodonium ion, 131
2,5-Dimethyltetramethyleneiodonium ion, 130
Di-(*m*-nitrophenyl)iodonium ion, 75
Dinorbornylchloronium ion, 26
Dinorbornylhalonium ions, 27
Diphenylamine, from phenylation of aniline, 72, 73
Diphenylbromonium ion, 67
Diphenyl ether, formation from diphenyliodonium hydroxide, 82
Diphenylhalonium ions, 6, 169
Diphenyliodine radical, 71, 73, 75
Diphenyliodonium bromide, 82
Diphenyliodonium 2-carboxylate, 74
Diphenyliodonium chloride, 64, 70, 75, 83, 84
Diphenyliodonium fluorobrate, 68, 72, 79, 80, 81
Diphenyliodonium halides, 80
Diphenyliodonium hydroxide, 82
Diphenyliodonium iodide, 63, 64, 68, 73, 78
Diphenyliodonium ion, 64, 66, 67, 69, 71, 72, 75, 76, 83, 86
Diphenyliodonium ion salt, 1, 68
Diphenyliodonium nitrate, 84
Diphenyl mercury, preparation from diphenyliodonium ion, 75, 80
Disproportionation, of dialkylhalonium ions, 9
Dissociation, of diiodonium ions, 68
  of polyiodonium ions, 68
2,2′-Dithienyliodonium ion, 58
η-Donor, alkylation with dialkylhalonium ions, 19
π-Donor, alkylation with dialkylhalonium ions, 19

Ether, alkylation of, 21
Ethylarylhalonium ions, 42
Ethylation, of toluene, 21
Ethylbenzene, in ethylation of benzene, 19
Ethylchlorocarbenium ions, 102
Ethylene, bromination of, 98
Ethylenebromonium ion, 4, 98, 115, 147, 148
Ethylenechloronium ion, 103
Ethylenehalonium ions, 98
  Carbon-13 nmr studies of, 113
  differentiation of π- and σ-bonded, 120
  1,2-dimethyl-, 103
  2,2-dimethyl-, 106
  2-ethyl-, 111
  parent ion, 98
  preparation via halogentaion of olefins, 120
  preparation via protonation of cyclopropyl halides, 118
  symmetrically substituted, 115
  tetramethyl-, 108
  unsymmetrically substituted, 116
Ethyleneiodonium ion, 98, 120

2-Ethy ethyleneiodonium ion, 111
Ethyl fluoroantimonate, 42
Ethyl α-phenylacetate, 93
Ethylphenylbromonium ions, 45
Ethyltoluene, 21

δ-Fluoroalkyl tosylates, 125
Fluoroantimonic acid, in preparation of dialkylhalonium ion, 8
1-Fluoro-2-bromoethane, 155
2-Fluoro-3-bromo-α-methylbutane, 107
2-Fluoro-1-bromopropane, 101
Fluorocarbenium ions, 109, 128
threo-2-Fluoro-3-chlorobutane, 106
1-Fluoro-2-chloroethane, 155
2-Fluoro-3-chloro-2-methyl butane, 107
1-Fluoro-4-iodobutane, 111, 128
2-Fluoro-3-iodo-2-methylbutane, 107
2-Fluoro-1-iodopropane, 101
p-Fluoromethyl phenyliodonium ion, 44
Fluoronium ions, 110, 130
  bridged, 110
5-Fluoro-*l*-pentyne, 135
2-Formyl-*l*-indanone, 71
Friedel Crafts reactions, role of dialkylhalonium ions in, 31, 45
Furan, ligand in halonium ions, 169
2-Furfuryl lithium, 59

Gallium bromide, in formation of dimethylbromonium ion, 32

2-Halo-3-acetoxy-2,3-dimethylbutanes, 110
β-Haloalkylcarbenium ions, 117
Halobenzenes, protonation of, 46
4-Halobenzenium ions, 6
1-Halo-3-bromopropanes, 140
Halocarbenium ions, 106
β-Haloethylarenes, 155
β-Haloethylation, 155
β-Haloethyl ethers, 155
1-Halo-2-fluoroethanes, 98
1-Halo-2-fluoromethylpropanes, 106
Halogenated tetromethylenehalonium ions, 134
Halogenation, of olefins, 120
Halogen atoms, anisotropy effect of, 15
  inductive effect of, 15
Halogen bridge, 1
Halogen complexes, differentiation of σ- and π-bonded, 120
  of alkenes, 120
Halogen participation, 1, 4, 125, 141
5-Halo-1-hexenes, 125, 129
1-Halo-3-iodopropanes, 140
2-Halo-2-methoxy-2,3-dimethylbutane, 110
5-Halo-2-methyl-2-pentene, 131
2-Halomethyltetramethylenehalonium ions, 134, 157
Halonium ions, 169
  acidic, 1, 4
  acyclic, 2, 5
  bicyclic, 4, 144
  classes of, 2
  cyclic, 147
  general aspects of, 1
  nomenclature of, 4
Halonium ylides, 90
5-Halo-1-pentynes, 125, 134, 135
Halophenium ions, 158
  heteroaromatic, 2, 158
Heats of formation, of cyclic halonium ions, 147
Heptamethylbenzenium bromoaluminate, 33
Heteroaromatic halonium ions, 2, 158
Heteroaryliodonium ions, 58
  reactions of, 86
Heteroorganic compounds, alkylation of, 21
Hexamethylbenzene, methylation of, 33
n-Hexylamine, reaction with diphenyliodonium ion, 75
tert-Hexyl cation, 24
n-Hexyl-*m*-nitrophenyl amine, 75
Hydridoalkylhalonium ions, 4
Hydridohalonium ions, alkyl, 6
  aryl, 6
2-Hydroxy-2-methyltetramethylenehalonium ions, 133

Influenza virus, biocidal effect by diaryliodonium ions, 89
"Iodine" tinctures, biocidal effect of, 89
Iodine(III)trifluoroacetate, 55
Iodoarenes, in preparation of diaryliodonium ions, 56, 57
Iodobenzene, in preparation of diphenyliodonium ion, 6, 39, 57, 61, 64, 68, 79, 80, 83, 84, 86, 91, 93, 94
2-iodobiphenyl, 167

2-Iodo-2′-biphenyl acetate, 167
p-Iododiphenyliodonium bisulfate, 54
2-(2′-Iodobiphenyl)diazonium fluoroborate, 165
4-Iodo-1-butyl tosylate, acetolysis of, 125
1-Iodo-2-fluorobutane, 111
erythro-2-Iodo-2-fluorobutane, 105
threo-dl-2-Iodo-2-fluorobutane, 105
1-Iodo-3-halobutanes, 111, 140
Iodomethane, in preparation of dimethyl-iodonium ion, 6
2-Iodo-2′-methoxybiphenyl, 167
erythro-dl-2-Iodo-3-methoxybutane, 106
threo-dl-2-Iodo-3-methoxybutane, 106
1-Iodo-2-methoxy-2-methylpropane, 107
Iodonium ions, 169. *See also* individual classes
  bearing on olefinic ligand, 59
  bearing on acetylenic ligand, 59
Iodonium ylides, 90, 91, 92, 93, 169
5-Iodo-1-pentene, 129
5-Iodo-1-pentyne, 135
Iodophene, 167
Iodophenium ions, 158
p-Iodophenyl bisulfate, 1
cis-2-Iodophenyl-1-methoxy-1-phenyl-1-hexene, 160
cis-2-iodophenyl-1-methoxy-1-phenyl-1-hexene, 160
trans-2-i(dophenyl-1-methoxy-1-phenyl-1-hexene, 160
Iodosoa ·enes, 55, 56
Iodosobenzene, 1, 54, 56
2-Iodosobiphenyl, 163
3-Iodosopyridine, 59
2-Iodothiophene, 86
Iodoxyarenes, 55
Iodyl sulfate, 55
Intermolecular exchange, of dialkylhalonium ions, 8, 23
Isobutylene, polymerization of, 31
Isoelectronic model compounds, 18
  for dialkylhalonium ions, 18, 19
Isopropyl bromide, exchange with bromonium ion, 24
Isopropyl cation, exchange with halonium ion, 23
Isopropyl chloride, exchange with chloronium ion, 24
Isopropyliodide, exchange with iodonium ion, 24
Isopropyl phenyl ketone, 71

Ketones, alkylation of, 21

trans-1-Lithio-2-o-lithiophenyl-1-phenyl-1-hexene, 158

Magnetic shielding effects, 17
Malonic esters, 69
Mercury, reaction with diaryliodonium ions, 80
Mesityl-*p*-tdyliodonium bisulfite, 85
Methanol, alkylation of, 156
Methylarylbromonium ions, 42
Methylation, of aromatics, 19
Methyl benzoate, alkylation of, 33
  formation from diphenyliodonium ion, 84
p-Methylbenzophenone oxime, phenylation of, 82
Methyl bromide, *see* Bromomethane
  effect on isobutylene polymerization, 31
Methyl chloride, solvent in isobutylene polymerization, 31
Methylchlorocarbenium ion, 101, 103
2-Methyl-5-chloro-2-pentanol, 130
Methylcyclopentyl cations, 30
o-(α-Methyl-β,β-dichloroethenyl)phenyl-diazonium fluorophosphate, 161
  thermal decomposition of, 162
1-Methyl-1,1-dimethylethylenehalonium ions, 116
2-Methylenetetramethylenehalonium ions, 134
2-Methylethylenebromonium ion, 116
Methyl fluoroantimonate, 39, 42, 142
Methylfluorocarbenium ion, 101
2-Methyl-5-halopentanol, 131
Methylhydridobromonium ion, 6
Methylhydridoiodonium ion, 6
Methyl iodide, *see* Iodomethane
  effect on isobutylene polymerization, 31
Methylmagnesium iodide, 71
Methylphenyliodonium ions, 7, 42, 45
2-Methylpropene, 31
Methy propyl ether, alkylation of, 156
2-Methyltetramethylenebromonium ion, 129, 143
2-Methyltetramethylenechloronium ion, 128, 138

3-Methyltetramethylenechloronium ion, 129
2-Methyltetramethylenehalonium ions, 128
2-Methyltetramethyleneiodonium ion, 129, 143
Molecular complexes, in halogen additions, 149

Nitrenes, halonium ylides in their preparation, 169
4-Nitrobenzoic acid, 67
Nitro compounds, alkylation of, 21
m-Nitrodiphenyliodonium tetrafluoroborates, 74
Nitromethane, alkylation of, 33
Nonclassical σ-delocalization, 29
Norbornene, bromination of, 153
Norbornyl cation, exchange with halonium ion, 27
2-*exo*-Norbornyl halide, exchange with halonium ion, 27

Olefins, *see* Alkenes
  preparation of their halonium ions, 120
Onium ions, 19, 21
Ortho effect, 21
Oxazole, 75, 77
Oxonium ions, 169

Pentamethylenehalonium ions, 2, 140
Perfluoroalkylaryliodonium ions, 39
Perfluoroalkyliodoso trifluoroacetate, 39
Pentafluorophenyllithium, 58
Pentamethylenebromonium ion, 128, 143
Pentamethylenehalonium ions, 142
Pentamethyleneiodonium ion, 143
Phenol, 79, 82, 83
  from diphenyliodonium ion, 79, 82, 83
Phenylating agents, 169
1-Phenyl-2-bromo-4,5-dicyanoimidazole, 93
1-Phenyl, 2-chloro-4,5-dicyanoimidazole, 93
Phenyl-*p*-chlorophenyliodonium chloride, 69
2-Phenyl-2-(α-chlorostyryl)-1,3-indandione, 72
Phenyl(α-chlorosylyl)iodonium fluoroborate, 72
β,β-Phenylchlorovinyl-phenyliodonium ion, 60
Phenyldiazonium salts, in preparation of diphenylhalonium ions, 87
5-Phenyl-5H-dibenziodole, 165, 166
p-Phenylenediamine, reaction with diphenyliodonium ion, 73
Phenylethynyllithium, 60
Phenylethynylphenyliodonium chloride, 60
2-Phenyl-1,3-indandione, 72
1-Phenyl-2-iodo-4.1.5-dicyanoimidazole, 94
Phenyliodoniobenzoic acids, 67
Phenyliodonium bisulfate, 1
Phenyliodosoacetate, 61, 62, 90, 94
Phenyliododichloride, 39, 56, 58, 60
Phenyllithium, 68, 69, 165
Phenylmercuric halides, 80
Phenylmercuric iodide, 88
Phenylmesityliodonium bromide, 85
Phenylpentafluorophenyliodonium ion, 58
2-Phenyl-2-phenylethynyl-1,3-indandione, 72
Phenyl(β-phenylethynyl)iodonium chloride, 72
Phenyl-2-thienyliodonium iodide, 86
Phenyl-2-thienyliodonium ion, 62, 86
Phenylti trichloride, 60
β-Phenylvinylmercuric bromide, 60
β-Phenylvinyltin trichloride, 60
Phosphines, reaction with diaryliodonium ions, 77
Phosphonium ions, 61, 78, 169
Phosphonium ylides, 61
Pinacolone dichloride, 110
Poly-2,6-dimethylphenylene ether, 82
Polyiodonium ions, 62
  chemical reactivity, 68
  degree of dissociation, 68
  kinetics of formation, 63
  IR spectra, 66
  NMR spectra, 64
  structural aspects, 63
  substituent effect of phenyliodonia group, 67
  UV spectra, 64
  X-ray studies, 63
Polymerization, of alkenes, 31
  of isobutylene, 31
Polymethylenehalonium ions, 4
Polystyryliodoso acetate, 63
Propadienylhalonium ions, 113
Propadienyliodonium ion, 113
Propionaldehyde, alkylation of, 33

Propylenebromonium ion, 101, 140
Propylenechlornium ion, 102, 103
Propyleneiodonium ion, 101, 140
Pyridine, ligand in halonium ions, 169
Pyridinium ylides, 91, 92
3-Pyridylaryliodonium ions, 59
2-Pyridyllithium, 59

Silver hexafluoroantimonate, in preparation of dialkylhalonium ions, 8
Solvolysis of, δ-chloroalkyltosylates, 125
δ-fluoroalkyltosylates, 125
Stability, of carbocations, 28
of cyclic halonium ions, 147
of halonium ions, 147
cis-Stilbene, 159, 160
trans-Stilbene, 159, 160
Sulfonylating agents, 44
Sulfonium ions, 169
Sulfoxylamine, 95

Tetrabenzoylethylene, 93
meso-1,2,3,4-Tetrachlorobutane, 134
2,3,5,6-Tetrafluorodibromobenzene, 52
2,3,5,6-Tetrafluoroiodobenzene, 52
Tetraiodonium ions, 62
2,3,5,6-Tetramethyl, dibromobenzene, 52
Tetramethylenebromonium ions, 38, 113, 125, 147, 148
Tetramethylenechloronium ion, 148
Tetramethylenehalonium ions, 4, 98, 125
Carbon-13 NMR studies, 135
2,5-dimethyl-, 128
halogenated-, 134
2-methyl-, 128
2-methylene-, 134
parent ion, 125
proton NMR study, 125
preparation via halogen participation, 125
symmetrically substituted, 135
tetramethyl, 133
trimethyl, 133
unsymmetrically substituted, 138
Tetramethyleneiodonium ion, 111, 128
2,2,5,5-Tetramethyltetramethylenechloronium ion, 133
Tetramethylethylene, 122
Tetramethylethylenebromonium ion, 115
Tetramethylethylenechloronium ion, 110
Tetramethylenethylenehalonium ions, 108
Tetraphenyliodophenium iodide, 158
2-Thienyl iodide, 86
2-Thienyllithium, 59
Thiophene, 58, 62, 169
2,5-Thiophenediyl-bis(phenyliodonium)ion, 62
Thiophene-2-sulfonate, 86
Titanium(III)chloride, 84
Toluene, 19, 63
ethylation of, 21
p-Toluenesulfonic acid, 92
o-Tolymesityl mesityl iodonium bromide, 85
Trialkyl phenylene triiodonium ions, 52
Trialkyloxonium ions, 21
Trialkylphosphate, 79
Triarylcarbenium ions, 1, 54
1,3,5-Tribromobenzene, 52
1,2,4-Tribromobutane, 134
2,4,6-Tribromomesitylene, 53
1,3,5-Tribromotrifluorobenzene, 53
Trichloroacetic acid, 91
1,2,4-Trichlorobutane, 134
1,1,2-Trichlorotrifluoroethane, 149
bromination in, 154
Triethylamine, 75, 81
Triethylphosphite, 78
Trifluoroacetic acid, 125, 134
Trihalonium ions, 52
1,2,5-Trihalopentanes, 134
1,3,5-Triiodobenzene, 52
2,4,6-Triiodomesitylene, 52
2,4,6-Trimethylacetophenone, 69
Trimethylenehalonium ions, 101, 140
Trimethyleneiodonium ion, 140
Trimethylethylenebromonium ion, 107
Trimethylethylenehalonium ions, 107
2,2,3-Trimethylethylenehalonium ion, 118
Trimethylethyleneiodonium ion, 107
Trimethylphenylenetriiodonium ions, 53
1,1,4-Trimethyltetramethylene carbenium ion, 139
2,2,5-Trimethyltetramethylene chloronium ion, 133
Trimethyltetramethylenehalonium ion, 133
Triphenyl boron, 165
Triphenyl iodine, 68
Triphenyl phosphine, 78
Tritylmagnesium chloride, 71

Trivalent organoidine intermediates, 159

Vinyl halides, 61
Vinylphenyliodonium halides, 61
Vinylphenyliodonium ions, 61
2-Vinyl-tetramethyleneiodonium ion, 135

Wagner-Meerwein rearrangement, 27
Water, alkylation of, 9